CAPRA OPTIMIZATION METHOD FOR SOLVING THE ECONOMIC DISPATCH PROBLEM WITH THE INTEGRATION OF WIND ENERGY SYSTEMS

AUGUSTEEN W A

CAPRA OPTIMIZATION METHOD FOR
SOLVING THE ECONOMIC DISPATCH
PROBLEM WITH THE INTEGRATION OF
WIND ENERGY SYSTEMS

First Edition September 2024

Written by AUGUSTEEN W A

TABLE OF CONTENTS

LIST OF TABLES

<table>
<tr><td>TABLE NO.</td><td>TITLE</td><td>PAGE NO.</td></tr>
</table>

LIST OF SYMBOLS AND ABBREVIATIONS

ABC	-	Ant Bee Colony Algorithm
π	-	Air density in kg/m^{-3}
	-	Angle suspended by the portion of unit circle
APSO	-	Anti-predatory PSO
PAPSO	-	Asynchronous Parallel PSO
BFA	-	Bacterial Foraging Algorithm
$x_{best,j}^{t}$	-	Best position of Capra i^{th} population
BGA	-	Binary Genetic Algorithm
$u_{i,j}^{t+1}$	-	Binomial recombination of jth vector of ith population
BBO	-	Biogeography-Based Optimization
ζ_{count}	-	Bite count of the Capra
ζ_{rate}	-	Bite rate of the Capra
$\zeta_{strength}$	-	Bite strength of the Capra
T_{h}	-	Bite time of the Capra in seconds
COA	-	Capra Optimization Algorithm
CMWE	-	Capture of Maximum Wind Energy
$[c_1...c_7]$	-	Characteristics constants of wind turbine
CSO	-	Civilized Swarm Optimization
CoDE	-	Composite DE
CF	-	Constant Frequency
CPSO	-	Cooperative Particle Swarm Optimization
x, y	-	Coordinates of Capra area
wc_{m}	-	Cost co-efficient of mth WES in \$/hr

$Fc_j(P_{tp,j})$ - Cost function of jth conventional generating unit in \$/hr

$F_{WES,m}(P_{WES,m})$ - Cost function of mth WES in \$/hr

CRX - Cross Over Value

CSOA - Cuckoo Search Optimization Algorithm

DE - Differential Evolution

DE/BBO - Differential Evolution / Biogeography-Based Optimization

DHS - Differential Harmony Search

DSM - Direct Search Method

DR_j - Down ramp-rate limit of jth conventional generating units in MW

ED - Economic Dispatch

EDWES - Economic Dispatch with Wind Energy System

EDWES - Economic Dispatch with Wind Energy System

$P_{tp,j}$ - Electrical power output of jth conventional generating unit in MW

$P_{WES,m}$ - Electrical power output of mth WES in MW

EAs - Evolutionary Algorithms

ESO - Evolutionary Search Optimization

$f(x_i)$ - Fitness function of ith population

$f(U_i^{t+1})$ - Fitness function of the feasible Binomial recombination vector of ith population

$f(X_i^t)$ - Fitness function of the feasible vector of ith population

$a_{tp,j}, b_{tp,j}, c_{tp,j}$ - Fuel cost co-efficient of jth conventional generating unit

Pg_w	-	Generated power in MW
GA	-	Genetic Algorithm
GAs	-	Genetic Algorithms
GB	-	Giga Bite
GHz	-	Giga Hertz
GW	-	Giga Watts
α	-	Grazing angle of Capra
	-	Grazing area of Capra
Ξ	-	Grazing land type
Hz	-	Hertz
HPSO-TVAC	-	Hierarchical Particle Swarm Optimizer with Time-Varying Acceleration Co-Efficient
HVDC	-	High Voltage Direct Current
HCS	-	Hill-Climb Search
HIP	-	Homogeneous Interior Point
HMF	-	Hopfield Modeling Framework
EP	-	Hopfield Neural Network
HNN	-	Hopfield Neural Network
$kw(1)$	-	Index co-efficient of mth WES
`	-	Indian currency in Rupee
$P_{tp,j}^{I}$	-	Initial output power of jth conventional generating units in MW of ith generation
$x_{i,j}$	-	Initial population of a vector.
KV	-	Kilo Volt
kg/m^3	-	Kilogram per cubic meter
LAM-CON	-	Lambda Consensus
LAM-ITR	-	Lambda Iteration
lr	-	Length of the rope

$P_{tp,j}^{\min}$ - Lower limit of jth conventional generating units in MW

MATLAB - Matrix Laboratory

$P_{tp,j,kz-1}^{U}$ - Maximum limit of jth conventional generating units prohibited zone in MW

$P_{tp,j,kz}^{L}$ - Maximum limit of jth conventional generating units prohibited zone in MW

$P_{tp,j,nzi}^{U}$ - Maximum limit of jth conventional generating units prohibited zone in MW

$x_{j}^{\max}$ - Maximum limit of jth individual vector

$t_{\max}$ - Maximum number of iteration

$C_{pw}^{\max}$ - Maximum power co-efficient

u_{m}^{mw} - Mean speed of the of mth WES in meter/seconds

MVA - Mega Volt Ampere

MW - Mega Watt

m/s - Meter per second

$P_{tp,j,1}^{L}$ - Minimum limit of jth conventional generating units prohibited zone in MW

MNRE - Ministry of Renewable Energy

MPSO - Modified Particle Swarm Optimization

MAPSO - Multi Agent system and PSO

$z_{i,j}^{t}$ - Mutation of jth vector of ith population

R_{1}, R_{2} - Mutually different integers that are also different from the running index i

NPSO - New PSO

NA - Not Available

ng - Number of conventional generating units

nzi	-	Number of prohibited operating zones
nw	-	Number of wind energy system in MW
ORCSA	-	One Rank Cuckoo Search Algorithm
ODE	-	Opposition-based DE
$C_{pw}^{optimal}\left(\lambda_{TSR}, \beta_{pitch}\right)$	-	Optimal power co-efficient
X_i^t	-	Parent vector of ith population
PSO	-	Particle Swarm Optimization
$\mu_{weq}\left(u_m^{mw}, T\right)$	-	per unit output under turbulence of mth WES in $/hr
β_{pitch}	-	Pitch angle
M	-	Population Size
$C_{pw}\left(\lambda_{TSR}, \beta_{pitch}\right)$	-	Power co-efficient
PSF	-	Power Signal Feedback
$P_{tp,j}^0$ -	-	Previous output power of jth conventional generating units in MW
$P_{tp,j}^{t-1}$	-	Previous output power of jth conventional generating units in MW of ith generation
POZ	-	Prohibited Operating Zones
PSO-MSAF	-	PSO with Modified Stochastic Acceleration Factor
QGA	-	Quantum Genetic Algorithm
r	-	Radius of the Capra's tether
R	-	Radius of the fence
R	-	Radius of the wind turbine
$P_{WES,m}^{rat}$	-	Rated power output of the mth WES in $/hr
RAM	-	Read Access Memory
RGA	-	Real coded Genetic Algorithm
N	-	Real-valued parameter vectors

U_i^{t+1}	-	Recombination vector of ith population
ω	-	Rotor speed in radian/second
$vw(l)$	-	Scale co-efficient of mth WES
X_i^{t+1}	-	Selection process j^{th} vector of ith population
SA	-	Self Adaptive
SA-EAs	-	Self Adaptive Evolutionary Algorithms
SARGA	-	Self adaptive Real coded Genetic Algorithm
SADE	-	Self-Adaptive Differential Evolution
$uw_m^{mw} / uw_m^{m\,rat}$	-	Short time duration average wind speed in p.u of jth WES
SDE	-	Shuffled frog DE
SiA	-	Simulated Annealing
SBX	-	Simulated Binary Crossover
SWT	-	Single Wind Turbine
SSA	-	Social Spider Algorithm
A	-	Surface area covered by the wind wheel
ε	-	Suspended angle between radius and unit circle
ℓ_{sward}	-	Sward height of the Capra in centimeters
11_{TSR}	-	Tip speed ratio
TSR	-	Tip-Speed Ratio
P_D	-	Total demand on the power system in MW
f	-	Total generation cost in \$/hr
TP	-	Total Power
TL	-	Transmission Loss
B_{ij}, B_{0i}, B_{00}	-	Transmission loss co-efficient
Pt_{loss}	-	Transmission loss in MW
	-	Turbulence Intensity in meter/seconds

T-NN	-	Two-phase optimization Neural Network
\$	-	United State currency in dollar
UR_j	-	Up ramp-rate limit of jth conventional generating units in MW
$P_{tp,j}^{max}$	-	Upper limit of jth conventional generating units in MW
VF	-	Variable Frequency
WECS	-	Wind Energy Conversion System
WES	-	Wind Energy System
Pw	-	Wind Power
v_w	-	Wind speed in meter/second

CHAPTER 1

INTRODUCTION

1.1   OVERVIEW

The consumption of electrical energy in developing countries like India is increasing exponentially on each successive day. The annual growth of energy demand in India has increased significantly from 304.5 GW in 2016 to 329.6 GW in 2017 (MNRE 2017). The conventional sources based power plants like thermal, nuclear, hydro and renewable sources based power plants like small-hydro, wind, bio-mass and solar plants are utilized to meet this demand. As on 31-03-2017, the total installed capacity of electrical power generation in India is 329204.53 MW. The conventional energy sources contribute 82.61% of the total power generation i.e., 271944.30 MW. The remaining 17.39% of power generation is contributed by renewable energy sources (57260.23MW) that includes small hydro-electric power plants (43798.66MW), wind power plants (32279.77MW), bio-mass plants (8311.78MW) and solar power plants (12288.83MW).

The energy demand in India is increasing day by day; As a result newer energy resources are beginning utilized to meet the required demand with the conventional sources of energy. Even though nuclear and renewable energy sources increases the power generation capacity it fails to meet steep increase in energy demand due to domestic and industrial consumption. So it is essential to maintain the optimal electrical power output from the available generation units. Therefore, in power system operation, optimization of generators' output and their scheduling has become essentially important. Ultimately, it leads to a dedicated field of study, Economic Dispatch (ED).

The main objective of ED problem is to minimize the total generation cost by meeting the specific demand satisfying the various constraints like power balance, generator power limits, prohibited operating zones and ramp-rate limits of each generator (Walter & Sheble 1993). A conventional thermal generator uses fossil fuels which are deplete-able in nature. Therefore, usage of these thermal generators should be reduced by increased usage of renewable energy resources. Among these renewable resources, Wind Energy System (WES) is extensively used that has significant advantages of relatively low investment cost and pollution free environment leading to wide usage of WES in recent years.

Conventional ED problem incorporating WES (EDWES) is one of the newer technologies that are being applied to meet the steadily growing electrical power demands. Apart from generation of electric power at minimized generation cost, EDWES also fulfils the generation of electrical power to meet the required specific energy demands and to operate generating units in feasible operating region (Sanjay Roy 2012). This EDWES plays a vital role in confronting many of the problems that are being faced in the energy demands of domestic and industries. Providing an effective solution to EDWES problem is constantly drawing the attention of the researchers involved in the field.

Apart from EDWES, efforts are being taken to ensure that the conventional thermal generators are operated more efficiently. Specifically, optimized dispatch of electrical power from conventional thermal generators is given due importance. The input-output characteristics of thermal power generators are piecewise-linear quadratic functions and by assuming the incremental cost curves of the thermal power generators as monotonically rising leads to inaccuracy in the resultant dispatch. Specifically, the input-output characteristics of these conventional thermal power generators are highly non-linear (Chiang 2005), because of the presence of non-linear constraints like generating unit limits, prohibited operating zones, and ramp-rate constraints (Kothari & Dhilon 2012) . The presence of these non-linear constraints converts the EDWES problem into a highly non-convex optimization problem, along with equality and inequality constraints resulting in determination of optimal dispatch as a challenging task. In addition, the complexity of EDWES problem is further increased by addition of the impact of turbulence of wind speed which makes it difficult for the conventional algorithm to provide solution for optimum dispatch (Sanjay Roy 2012).

Capture of Maximum Wind Energy (CMWE) from a wind turbine is another non-linear problem which is addressed in this thesis. Conventionally, the extracted energy from wind is converted into electric energy using a squirrel cage induction generator or doubly fed induction generator and is supplied to the grid or a standalone load. The main disadvantage of this arrangement is its poor efficiency as it does not track the maximum power from wind turbine as the velocity of wind changes (Oriol Gomis-Bellmunt et al. 2010). Therefore, the optimization of wind turbines for CMWE has gained greater importance in recent research in power system operation.

1.2   SOLUTION METHODOLOGIES

Many traditional optimization methods are employed to solve the ED, EDWES and CMWE problems. These traditional methods are unable to solve efficiently the ED, EDWES and CMWE problems that involves non-continuous, non-linear and non-convex solution space that are associated with large scale power systems.

However, with advancement in population based heuristic optimization algorithms leads to the development of Evolutionary Algorithms (EAs). EAs uses mechanisms inspired by biological evolution (mutation, recombination, selection) which is known as population based techniques that explore the search space randomly by using several individual solutions. EAs are capable of providing an alternative for obtaining better optimal solutions especially for the problems with non-continuous solution spaces. EAs exploit the solution in the feasible region based on random exploration instead of exploring entire search region resulting in faster convergence of the optimal solution with lesser computational resources.

In the field of evolutionary computation, popular methods that have been employed are Genetic Algorithm (GA) (Po-Hung et al. 1995), Particle Swarm Optimization (PSO) (Gaing 2003) and Differential Evolution (DE) (Nasimul & Hitoshi Iba 2008). These methods provide better convergence characteristics and have an ability to find global optimal solution. Hence these methods are widely used by the researchers for their study.

GA is a heuristic search algorithm that combines Darwinian survival of the fittest principle with genetic operation (Goldberg 1989), to form an effective method for finding optimal solutions. GA determines the optimal solution by using three basic procedures: 1) reproduction, 2) crossover and 3) mutation. In reproduction process the selection of trail vector for the next generation is performed based on their best fitness values. In crossover process new offspring are created by selecting randomly from the mating pool of individual strings. In mating pool a crossover site is selected at random along the string length of the binary digits that are inter-changed between two strings to form a new offspring. The mutation process creates new offspring with random alternation of the binary digit which inserts a new parameter to the current population in the optimization process. This process is repeated for several generations for the evaluation till the solution moves toward the optimal value.

PSO has become increasingly popular optimization problem in the field of evolutionary computation. It is based on bio-inspired behaviour of swarm as fish schooling and bird flocking. PSO determines the optimal solution by using two basic operations: position update of the particle and velocity update of the particle. Position update and velocity update of the particles is being done according to its own experience and the experience of neighbouring particles, thereby making use of best positions encountered by itself and its neighbours (Kennedy & Eberhart 1995). PSO method has been successfully applied to solve large scale power system optimization problems.

Recently, Differential Evolution (DE) has become popular optimization method for solving power system problems. DE is a population based stochastic evolutionary computation method and it differs from conventional GAs in terms of perturbing vectors (Storn & Price 1997) which are the difference between two randomly chosen parameter vectors. DE determines the optimal solution by combining simple arithmetic operators with classical operators of recombination, mutation and selection from a randomly generated starting population to a final solution.

The optimal solution of ED, EDWES and CMWE problem are solved by many optimization methods. Predominantly EAs like GA, PSO and DE are widely applied now-a-days for solving power system problems. These optimization methods undergo heavy computational burdens which leads to premature convergence and as a result the performance of these algorithms are affected when applied to the problem with multiple local optimal solutions. Additional constraints handling capacity need to be enhanced that are associated with practical power system problems. Another issue which demands attention is determination of a termination criterion. The terminating criterion must be such that the global optimal solution should be obtained with less computational burden. In order to overcome the above issues, attempts have been made to meliorate optimization methods, develop new methods and employ it for better solution.

A new bio-inspired Capra Optimization Algorithm (COA) is proposed in this thesis and it has been successfully employed to solve the ED, EDWES and CMWE problems for the considered test systems.

The proposed COA has the power of exploration and exploitation by utilizing Capra parameters, namely grazing area, length of the rope and bite rate. The above parameters help the COA in attaining better convergence towards optimal solutions. The performance of COA is investigated on ED, EDWES and CMWE problems for standard test systems and results are compared with other existing / recently reported algorithms in the literatures.

1.3   ORGANIZATION OF THE THESIS

The contents of this thesis are likely to be covered by the contributions given in the list of publications.

This thesis is organised in seven chapters as described as follows:

Chapter 1, "Introduction" provides an overview of the significance of ED, EDWES and CMWE problem. The different optimization methods that are applied to solve these problems are discussed. In addition to conventional ED problems, issues involved in incorporating WES to the conventional ED problems and capture of maximum power from WES using the evolutionary optimization methods are discussed in elaborately.

Chapter 2, "Literature Survey" provides the technical background of the thesis and reviews of the existing optimization methods for solving the conventional ED problem without valve point loading and non-linear constraints like transmission losses, prohibited operating zones, and ramp rate limits. In addition, the essence of incorporating WES to the conventional ED problem is addressed and also different methods employed in solving EDWES problem are reviewed. Also, this chapter discusses the problem of capturing maximum power extraction from WES and reviews the methods in solving CMWE problem.

Chapter 3, "Economic dispatch problem with, without WES and Capture of maximum power from WES" presents the formulation of conventional ED problem without valve point loading and the constraints that are considered are described and detailed formulation of EDWES problem for short duration wind speed is described in this chapter. Also detailed formulation of "Capture of maximum power from WES" describes the problem formulation for CMWE for constant frequency off shore wind farm and its constraints involved are explained in this chapter.

Chapter 4, "Modeling of bio-inspired Capra Optimization algorithm" deals with the theoretical development and modeling of newer bio-inspired COA in order to prevail over non-convex and non-linear power system problems.

Chapter 5, "Results and Discussion" presents the simulation result obtained by implementing the proposed COA to ED, EDWES and CMWE problems. A theoretical validation of COA algorithm is also explained in this chapter along with the implementation of proposed COA algorithm to ED, EDWES and CMWE problems with various test systems. The effectiveness and efficiency of the proposed COA algorithm is highlighted by comparing the results obtained from other well-known existing/recent optimization methods.

Chapter 6, "Conclusion and Scope for Future work" summarizes the outcome of the proposed research work and outlines the possible development for the future work.

CHAPTER 2

LITERATURE SURVEY

2.1   INTRODUCTION

Day by day research is being advanced in the area of soft computing and which is used to solve power system problems. Many research and technical articles have been published by the researchers using soft computing methods to solve ED, EDWES and CMWE.

This chapter describes a detailed survey of research works that have been carried out using conventional and EAs (Bansal 2005). Further, EAs methods proposed in literature like GAs, PSO, and DE methods are described in detail. In addition this chapter specifically focuses on the various methods that have been employed to solve ED, EDWES and CMWE problems. The motivation of this research work and the objective of this thesis are described in detail in this chapter.

2.2   CONVENTIONAL OPTIMIZATION METHODS

Conventional optimization methods have been used over the past few decades for power system problems. The following four methods are the most popular conventional methods.

Lambda iteration method (Lin & Viviani 1984) to solve ED problem uses piecewise quadratic cost function multiplier called Lagrange multiplier (Lambda). In this method Lambda has been calculated by solving systems of equation and the inequality constraints are to be fulfilled in each trial the equations are solved by the iterative method.

Linear programming (Wood et al. 1996) is the optimization method which is used to solve linear objective function. In this method problem is tackled by expressing the non-linear input-output equations as a set of linear functions.

Dynamic programming (Wood et al. 1996) considers optimization problem as an allocation problem. In this approach single optimum solution is not calculated rather it generates a set of discrete points to obtain optimal solution.

Integer and mixed-integer programming (Bansal 2005) uses the decomposition technique to decompose continuous problem to integer programming. When the objective function has independent variables that can take only integer values such objective functions are solved by integer programming and when some of the independent variables are continuous the problem is solved by mixed integer programming.

The classical methods employ mathematical programming to solve objective function but when the objective function has non-linear characteristics these classical methods fail to achieve optimal solution satisfactorily. In addition, these classical methods find it difficult to solve the optimization problem with equality and inequality constraints. However, dynamic programming method can solve the problem with this non-linearity but it suffers from curse of dimensionality or local optimality (Nidul Sinha et al. 2003). The evolution of intelligent techniques has provided global optimum solution or near solution overcoming the difficulties faced by conventional optimization methods. Specifically EAs is emerged as the most popular method which is successfully employed to the optimization problems.

2.3 EVOLUTIONARY ALGORITHMS

Evolutionary algorithms are successfully employed for solving conventional ED problem having with multi modalities, non-linearities and non -convex. The EAs used for obtaining optimal solutions are described as follows,

2.3.1 Genetic Algorithm

Walter & Sheble (1993) presented Genetic Algorithm (GA) to solve ED problem. GA is search method inspired from the natural selection, survival of fittest and natural genetics. This method obtains optimal solution by combining randomized structured exchanges of information between the solutions.

Deb & Agarwal (1995) improved the solution quality of GA in terms of probability of creating an offspring solution from a given parent solutions and also introduced a cross over operator for Real-coded Genetic Algorithm (RGA) with Simulated Binary Crossover (SBX). These modifications in GAs lead to perform better with single point crossover. Moreover, GA with SBX is found to be useful in problems having multiple optimal solutions.

Deb & Goyal (1996) suggested genetic adaptive search by combining genetic adaptive search method and mixed functional objective function. This method is the modified form of the standard GA thereby reducing convergence burden to determine optimal solution. Beyer & Deb (2001) proposed real-parameter cross-over operators and self-adaptive evolution strategies as the properties of Self-Adaptive Evolutionary Algorithms (SA-EAs) for successful in real valued search spaces. Especially the population mean and variances of a real number lead to similar performance in self adaptive GAs and EAs.

Subbaraj & Rajnarayanan (2009) proposed an improved Real-Coded GA (RGA) with Self Adaptive (SA) phenomenon and improved RGA is called SARGA. SARGA has been successful in solving problems with non-linearities.

2.3.2   Particle Swarm Optimization

Kennedy & Eberhart (1995) proposed PSO, an optimization technique to solve non-linear objective functions. The algorithm is based on social behaviour of swarm and it is a population based search approach for determining optimal solution.

Shi & Eberhart (1998) suggested a new parameter called inertia weight that incorporated in the PSO to improve the performance of PSO on search space to find optimum solution.

Ratnaweera et al. (2004) introduced a parameter called automation strategy for the PSO. To control the local search and for effective convergence of PSO, time-varying acceleration co-efficient is introduced apart from time-varying inertia weight factor. To improve the performance of PSO, a hierarchical particle swarm optimizer with time-varying acceleration co-efficient has been proposed. Berg F et al (2004) proposed cooperative PSO

(CPSO) that includes a parameter called cooperative behaviour. This cooperative behaviour is achieved by using many swarms to optimize various components of the solution vector cooperatively.

Zhao et al. (2005) proposed a method that integrates multi agent system and PSO (MAPSO). In MAPSO an agent represent a particle to PSO and it represents the candidate solutions to the optimization. To improve the effectiveness of PSO, each agent competes and cooperates with its neighbour.

Immanuel & Thanushkodi (2008) proposed Anti-predatory PSO (APSO). Anti-predatory activity of the bird is introduced as a parameter in APSO and the inclusion of this parameter increased the performance of PSO

Subbaraj et al. (2010) introduced the PSO with Modified Stochastic Acceleration Factor (PSO-MSAF). In this method PSO is modified with parallel computing where multiple machines are employed simultaneously to solve optimization problem in large scale power systems. The extra diversification capability provided by the modified stochastic acceleration factor improves the performance. Since this method balances both exploration and exploitation, the chances for the solution to be trapped in a local optimum are minimized.

2.3.3   Differential Evolution Algorithms

Storn & Price (1997) introduced Differential Evolution (DE) algorithm utilizing three basic operations, mutation, crossover and selection in order to reach an optimal solution. In DE all variables are changed together during the crossover operation and in preliminary stage of search the solutions moves very fast towards the optimal solution point.

Fan & Lampinen (2003) modified the DE algorithm with a local search operation parameter and trigonometric mutation. The modified DE algorithm has a better convergence and consequently determines the good sub-optimal solution within the short duration of time.

Shahrya et al. (2008) introduced Opposition-based DE (ODE) algorithm. This ODE employed opposition based learning for the initialization of the population and also for generation jumping. ODE improved the solution quality by using the parameter opposite numbers.

Pedersen (2010) proposed a meta-optimized parameter for DE that selects the good choices of parameters for various optimization scenarios to improve the performance of DE algorithm.

Mallipeddi et al. (2011) proposed ensemble of mutation strategies and control parameters for DE. In this method a distinct mutation strategies along with a set of values for each control parameters of DE coexists throughout the evolution process and competes to produce the better optimum solution. Yong et al. (2011) introduced a method known as Composite DE (CoDE) containing three vector generation strategies and control parameter settings. The performance of this method is influenced by trial vector generation strategy and control parameters.

2.4   CONVENTIONAL ED PROBLEM WITHOUT WES

ED plays a vital role in the smooth functioning of power system operation and control. The main aim of ED is to allocate the optimal generation of power among the available generating units that satisfies the constraints and minimizes the generation cost. In the past, solution of ED problem is obtained by analytical or numerical computation for small scale power systems. But now, because of large scale power systems with discontinuous non-convex and non-linear objective function, solution of ED problem by conventional optimization methods is difficult. Hence there is a need for a suitable optimization algorithm to solve the ED problem.

Lee & Breipohl (1993) introduced a method for solving the ED problem with reserve constraint of the generating units having prohibited operating zones. This prohibited operating zone decomposes and divides the operating region between minimum and maximum of operating generating units limit and used Lagrange relaxation method to solve this ED problem with prohibited operating zones.

Park et al. (1993) described Hopfield Neural Network (HNN) which solve ED problem as piecewise quadratic cost function. In this HNN multiple intersection cost functions were used for each generating unit and also calculated the transmission losses for obtaining the optimal solution. Walter & Sheble (1993) solved an ED problem with valve-point discontinuities and transmission loss using two different encoded techniques of GAs.

Po-Hung et al. (1995) proposed a GA for solving ED problem in large scale power system. In this approach, GA used stochastic operators instead of deterministic rules to search for the optimal solution. Moreover, the approach considered network losses, ramp-rate limits, and prohibited zones to solve the ED problem.

Yang et al. (1996) developed Evolutionary Programming (EP) approach to solve ED problem with non-smooth fuel cost functions. In this

EP, optimal solution is determined based on the current trail solution and selection based on successive generation and it is applied to the system with transmission losses for Taipower system.

Rabih et al. (2000) presented a Homogeneous Interior Point (HIP) method for the ED problem that combines both independent blocks of constraints and coupling constraints into a single optimization problem. This method reduces the ED problem into a convex optimization problem that possesses non-linear inequality constraints and free variables with network constraints and transmission losses.

Su & Lin (2000) proposed Hopfield Modeling Framework (HMF) to ED problem. In this HMF model an energy function composed of transmission loss, total fuel cost and power mismatch was defined to find an optimal solution. The factor weight related with energy function is estimated directly.

Yalcinoz et al. (2001) developed a GA approach based on arithmetic crossover for solving the ED problem including transmission losses. Elitism, arithmetic crossover and mutation are used in GA to generate successive sets of possible operating solutions.

Gaing (2003) employed PSO algorithm for solving the ED problem considering generator constraints. Many non-linear characteristics of the generator such as transmission losses, prohibited operating zone, ramp-rate limits and non-smooth cost functions are considered in this algorithm. Nidul Sinha et al. (2003) presented an EP method and used the control parameters as real values, but not as binary codes as in traditional GAs. This EP relies primarily on mutation and selection, but not crossover, as in traditional GAs. Hence, considerable computation time has been successfully saved by this EP.

Naresh & Dubey (2004) proposed Two-phase optimisation Neural Network (TNN) modelling framework for solving the ED problem in large-scale power systems with network losses, ramp-rate limits and prohibited operating zones. This TNN method employed a set of differential equations obtained from transformation of Lagrangian energy function. Aruldoss Albert & Ebenezer Jeyakumar (2004) proposed integrating the PSO with sequel quadratic programming technique for solving large-scale system to solve ED problem.

Immanuel & Thanushkodi (2007) proposed New PSO (NPSO) to solve non-convex ED problem with network losses. In this NPSO, the particle remembers its best position previously visited by its cognitive behaviour that helped to improve the performance of this NPSO to solve non-convex ED problems. Immanuel & Thanushkodi (2008) proposed the Anti-Predatory PSO (APSO). APSO is based on the anti-predatory behaviour of the bird. Immanuel & Thansushkodi (2009) also proposed Civilized Swarm Optimization (CSO) based on modifying the parameter in PSO exploiting to solve ED problem including transmission losses.

Nasimul & Hitoshi Iba (2008) employed DE algorithm with specially measured equality and inequality constraints which solves the real world problems applied to generating units including transmission loss.

Subbaraj et al. (2009) developed SARGA based on tournament selection along with simulated binary crossover. The selection process is most effective in exploration capability by creating tournaments between two solutions. Subbaraj et al. (2010) Proposed Parallelized PSO with Modified Stochastic Acceleration Factor (PSO-MSAF) for large scale power systems. In this PSO-MSAF random cognitive and social learning parameter are introduced into a parallel computing environment for better convergence to optimal solution.

Aniruddha & Chattopadhyay (2010) proposed Biogeography-Based Optimization (BBO) algorithm for ED problems. BBO dealt with the geographical distribution of biological species. In this method optimization problem is described from behaviour of how a species can arise, migrate from one habitat to another. However, BBO has some features that are in common in GAs and PSO. Nima & Hossein (2010) proposed a modified DE with a new mutation operator and selection mechanism inspired from GA, PSO and simulated annealing (SiA).

Sayyid et al. (2012) proposed a gumption approach for ED problems with losses. This method is not heuristic but it enhances the convergence faster and it is successfully applied to the ED problem including losses.

Basu & Chowdhury (2013) proposed bio-inspired algorithm Cuckoo Search Optimization Algorithm (CSOA). CSOA is based on the behaviour of obligated brood parasitism of cuckoo and it is successfully employed to ED problem with transmission losses. Srinivasa reddy & Vaisakh (2013) proposed a Shuffled frog DE (SDE) for large scale ED problems and it is based on the benefits of shuffled frog leaping behaviour with DE algorithms.

James & Victor (2016) proposed a Social Spider Algorithm (SSA) for solving ED problems including transmission losses. The SSA method is developed from mimicking and foraging behaviour of social spiders. This method explores and exploits the solution search space by iterative manner.

2.5 CONVENTIONAL ED PROBLEM INCORPORATING WES

The various methodologies and ideas used for solving conventional ED problem incorporating WES reported in literature is discussed below.

Chen et al. (2006) proposed a Direct Search Method (DSM) approach to wind-thermal energy coordination. In this approach a better wind-thermal coordination is developed to determine the optimal allocation of wind generator capacity that can be integrated into conventional thermal generating system reliably and efficiently.

Hetzer et al. (2008) described the Wind Energy Conversion System (WECS) to conventional ED problem by developing a model for including WECS to ED problem. In addition it also accounted the cost factors associated to over-estimation and under-estimation of available wind power. In this approach stochastic wind is characterized as Weibull probability density function.

Farhat & El-Hawary (2010) introduced dynamic adaptive Bacterial Foraging Algorithm (BFA). BFA method is based on the behaviour of bacteria E-Coli that helps in solving ED problem incorporating WES.

Jiang et al. (2011) implemented PSO for solving the ED problem with WES and this approach combines a solution sharing strategy with an elitist learning strategy. Lee et al. (2011) proposed Quantum Genetic Algorithm (QGA) for solving ED problem including wind power system.

Sanjay Roy (2012) proposed short duration wind variation in which randomness is introduced in wind speed around a short-duration stable mean wind speed. It investigates the effect of wind energy based generating capacity incorporated with ED problem and also introduced a cost co-efficient for WES to short duration variations of wind speed.

Mahshid & Mehdi (2013) proposed ED problems incorporating wind power plant and obtained optimum solution using Modified Particle Swarm Optimization (MPSO) method. In this proposed approach the total cost is dependent on wind speed during a specific period of time. Benhamida et al. (2013) presented an approach for ED problem incorporating WES using PSO and also included the short time duration wind speed variations as a static power instead of stochastic model of WES.

2.6   CAPTURE OF MAXIMUM WIND ENERGY

The various methodologies and ideas used for capturing maximum wind energy from a wind farm reported in literature is discussed below.

El-Khazendar & Ahmed (1994) proposed a methodology to capture maximum wind energy by designing a cheap, maintenance-free generator called switched reluctance generating system.

Hui et al. (2004) proposed a neural network based small wind generation system for maximum wind power extraction. Neural network is used for wind speed estimation and to determine power co-efficient of wind turbine. Andrew kusiak & Song (2010) designed a wind farm layout for

CMWE. The designed model included wake loss and direction of wind of WES. EA with bi-criteria optimization is utilized to solve the designed model.

Shuhui Li et al. (2010) investigated CMWE using look-up table approach. In this approach peak power-tracking scheme is implemented based on direct-current vector control and it also used advanced power electronic technologies for CMWE problem. Oriol Gomis-Bellmunt et al. (2010) dealt CMWE using PSO algorithm and evaluated the power generated by variable and constant frequency offshore wind farms connected to a single large power converter. A methodology to analyze different wind speed scenarios and system electrical frequencies is presented and applied to the test system considered.

Zivkovic et al. (2012) proposed a fuzzy logic controller for wind turbine control. Wind turbine control loop provides the reference inputs for the electric generator in order to make the system run with maximum power. A separate wind speed estimator is also proposed since the wind speed involved in the aerodynamic equations is a stochastic variable and its effective value cannot be measured directly. Souma (2012) proposed a method for CMWE using unrestricted wind farm layout optimization. This method determines the optimum farm layout and correct selection of turbines for CMWE.

Yingming Liu (2013) discusses the control strategy of doubly-fed wind generator to capture the maximum wind energy. Due to random fluctuations of the wind speed, the speed of the doubly-fed wind generator has to be adjusted constantly to track the maximum power of wind generator by maintaining optimal tip speed ratio. Mirhamed Mola (2013) analyzed CMWE under various environmental conditions by considering both low-wind and high-wind circumstances using fuzzy logic controller.

Gebraad & Wingerden (2014) presents a data-driven adaptive scheme to adjust the control settings of each wind turbine in a wind farm such that an increase in the total power production of the wind farm is achieved. This is carried out by taking into account the interaction between the turbines through wake effects. The optimization scheme is designed in such a way that it yields fast convergence so that it can adapt to changing wind conditions quickly.

2.7   MOTIVATION OF THE THESIS

The research works of various methods were briefed in previous sections, from this it can understood that, the evolutionary based optimization methods are much suited for solving ED, EDWES and CMWE problems when compared to traditional techniques. However, the evolutionary methods suffer from premature convergence and long execution times resulting slower convergence of optimization problem. However, in order to overcome the above drawbacks, many attempts have been made and to prevent early convergence researches are in still under development.

Therefore, this thesis engage in developing a new bio-inspired algorithm to implement to the conventional ED, EDWES and CMWE problems thereby enhancing the solution quality when compared to the other well-known and most popular optimization methods like GA, PSO and DE by improving the computational time.

2.8   OBJECTIVE OF THE THESIS

This thesis, figure out the performance and competence of the existing optimization techniques in finding the optimal solution of ED, EDWES and CMWE problems. This thesis proposes a bio-inspired optimization algorithm known as Capra Optimization Algorithm (COA) to enhance solution quality, constraint handling and optimality verification without affecting the quality of the solution.

The specific objectives of this thesis work are:

1. Theoretical development and modeling of bio-inspired Capra Optimization Algorithm (COA) in order to overcome and enhance solution quality in non-convex power system ED problems.

2. Implementation of COA technique to obtain the optimal solution for conventional ED problem considering power balance prohibited operating zones, ramp-rate limits and transmission losses as constraints.

3. Effectiveness of COA technique has been analyzed by implementing it to EDWES problem.

4. COA technique is implemented for CMWE problem for constant frequency off shore wind farm.

The contents of the above research work contributions are expected to be covered in the list of publications.

2.9 CONCLUSION

Different research works employed relating to the optimum solution to ED, EDWES and CMWE. The different EAs and their development, enhancement in solving ED, EDWES and CMWE problems and their limitation are discussed. In addition, the motivation and the objective of this thesis are discussed.

CHAPTER 3

ECONOMIC DISPATCH PROBLEM WITH, WITHOUT WIND ENERGY SYSTEM AND CAPTURE OF MAXIMUM POWER FROM WIND ENERGY SYSTEM

3.1   ECONOMIC DISPATCH PROBLEM WITH, WITHOUT WIND ENERGY SYSTEM

Recently, the electric power industry has undergone considerable changes in operation and control due to the ever-increasing electrical requirements of the growing large population and rapid growing industrial sector. Increasing demand and decreasing energy sources such as fossil fuels necessitates the optimum usage of available resources. Hence the power system operator not only has to meet the load demand but also needs to operate the power system in the most economic way. ED problem in power system is to realize the optimal allocation of generating units to meet the power demand while satisfying the operational constraints.

Section 3.2 formulates ED problem of conventional power plants without incorporating wind energy system (WES). The constraints associated with the conventional ED problem are: bounded operational limits; network transmission loss, Prohibited Operating Zones (POZ) and ramp-rate limits which are explained in this section. Section 3.3 deals with the formulation of ED problem incorporating WES (EDWES). Also, this section briefs the generation of wind power from WES for a short duration wind speed involving the effect of turbulence.

3.2 CONVENTIONAL ED PROBLEM WITHOUT INCORPORATING WES

Since the electrical power cannot be stored, the electrical power is produced from natural sources and is delivered to the load as and when the demand rises. A system which is used for the delivering the bulk power for significant long distances is called as transmission system and the system which is used for local delivery of power is called as distribution system. Hence, an interconnected power system as shown in Figure 3.1 consists of three stages: generators which generate electrical energy, transmission lines for transmitting power to large distances and the loads which make use of the generated power. (Wood & Wollenberg 1996)

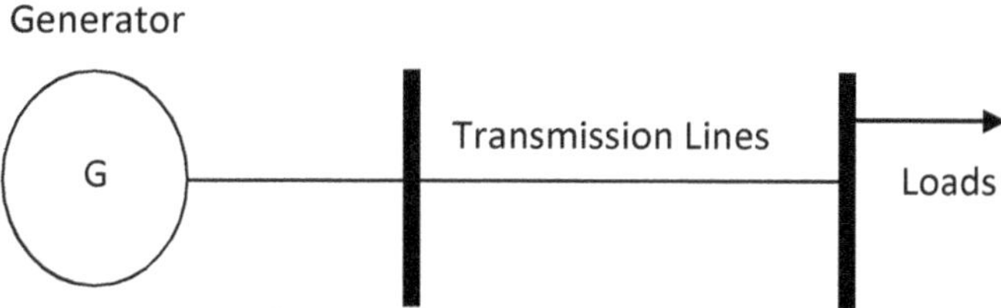

Figure 3.1 A simplified interconnected power system

Various sources of generating electrical power are: coal, oil or gas, river water, marine tide, a radioactive material, sun, wind and so on. Since there exist numerous sources, the picking of a particular source of energy is based on economical, technical or geographic basis. But, in general, in an interconnected system, the electrical power from different sources is used to meet the demand such that the operating expenses of generating units are minimized.

Conventional ED problem is one of the techniques used in interconnected system to meet the scheduled loads. Hence, the objective of the ED problem is to determine the optimal generation level of each available generating unit, so that the total cost of the generation is minimum for a given scheduled loads by satisfying all the constraints. Mathematically, the conventional ED problem can be formulated as (Su & Lin 2000).

$$\min f = \sum_{j=1}^{ng} Fc_j\left(P_{tp,j}\right) \tag{3.1}$$

where,

$$Fc_j\left(P_{tp,j}\right) = \sum_{j=1}^{ng} a_{tp,j}\, P_{tp,j}^2 + b_{tp,j}\, P_{tp,j} + c_{tp,j} \tag{3.1a}$$

The different constraints to be considered for the conventional ED problem are: equality and inequality constraints which are discussed below:

3.2.1 Equality Constraint

The essential equality constraint is the power balance constraints. The total electrical power generated from the available generating units has to meet the specified power demand and the transmission losses which is mathematically expressed as: (Su & Lin 2000).

$$\sum_{j=1}^{ng} P_{tp,j} = P_D + Pt_{loss} \tag{3.2}$$

where

$$Pt_{loss} = \sum_{i=1}^{ng}\sum_{j=1}^{ng} P_{tp,i}\, B_{ij}\, P_{tp,j} + \sum_{i=1}^{ng} B_{0i}\, P_{tp,i} + B_{00} \tag{3.2a}$$

3.2.2   Inequality Constraint

3.2.2.1   Bounded real power limits

The electrical power generated from each available generator has to be within its maximum and minimum generating limits and is given by

$$P_{tp,j}^{\min} \} P_{tp,j} \} P_{tp,j}^{\max} \tag{3.3}$$

3.2.2.2   Ramp-rate limit

Normally, the assumption of adjusting the unit generation output of each available generator instantaneously is used in conventional ED problem. Although, this assumption simplifies the problem, it does not replicate the actual processes happening in each available generating unit. Hence, to predict the actual operating process, the inequality constraint of each available generator is limited using ramp - rate limits. Three different possible operating situations of each unit which is available from $t-1$ to t are: (1) steady-state operating limits, (2) increase in power generation and (3) decrease in power generation which is shown in Figure 3.2 to Figure 3.4 respectively.

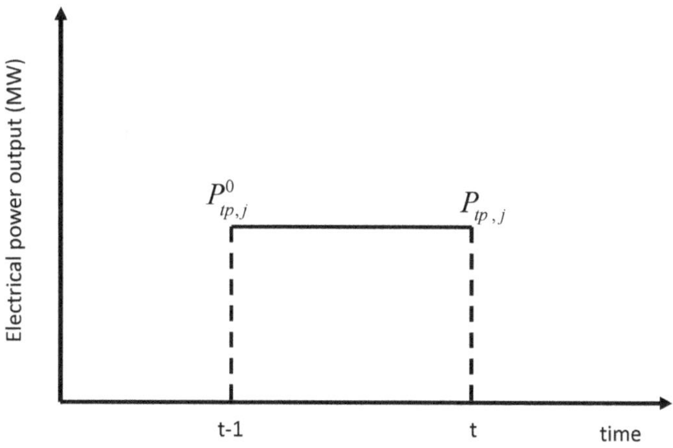

Figure 3.2 Steady state operation of an on-line unit

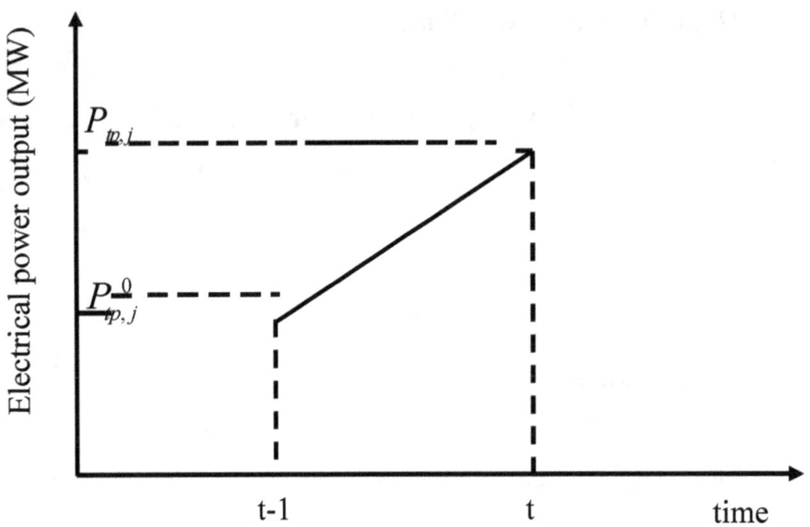

Figure 3.3 Increasing power output of an on-line unit

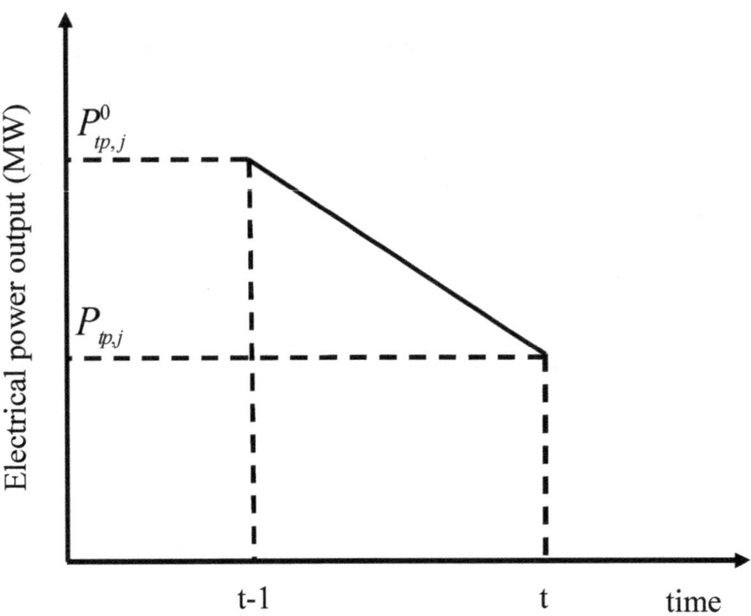

Figure 3.4 Decreasing power output of an on-line unit

Therefore, the inequality constraints of each generator based on ramp-rate limits are:

(i) if generation decreases:

$$P_{tp,j}^{t-1} - P_{tp,j}^{t} \leq DR_j$$

(ii) if generation increases:

$$P_{tp,j}^{t} - P_{tp,j}^{t-1} \leq UR_j$$

Combining the above two equations, the ramp-rate limit constraint of each available generator is given below:

$$\max\left(P_{tp,j}^{\min}, P_{tp,j}^{0} - DR_j\right) \leq P_{tp,j} \leq \min\left(P_{tp,j}^{\max}, P_{tp,j}^{0} + UR_j\right) \tag{3.4}$$

3.2.3 Prohibited Operating Zone

Typical generating units possibly will have vibrations in a shaft bearing caused by a steam valve develops prohibited operating zone which converts the continuous cost function into a discontinuous cost function as shown in Figure 3.5. Also, using real performance testing or operational records, it is complicated to understand the cost function in these prohibited operating zones. In real practice, the best economy is achieved by avoiding the operation of generators in these areas.

The electrical power output of a generating unit is required to avoid all capacity limits and unit operation in prohibited operating zones. The existence of prohibited operating zones is taken into account by the following constraints in conventional ED problem.

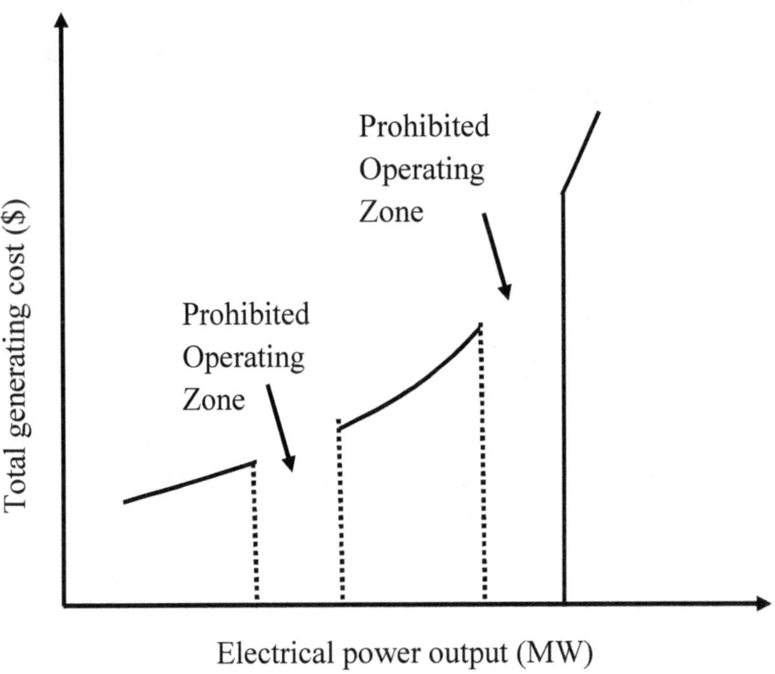

Figure 3.5 Input-output characteristics with prohibited operating zones

$$P_{tp,j}^{min} \succ P_{tp,j} \quad \succ P_{tp,j,1}^{L} \tag{3.5}$$

$$P_{tp,j,kz-1}^{U} \succ P_{tpj} \quad \succ P_{tp,j,kz}^{L} \qquad kz = 2,3,...,nzi \tag{3.6}$$

$$P_{tp,j,nzi}^{U} \succ P_{tpj} \quad \succ P_{tp,j}^{max} \tag{3.7}$$

3.3 CONVENTIONAL ED PROBLEM INCORPORATING WES

EDWES is one of the newer technologies in the power system. In the midst of the on-going search for alternative to typical energy resources, with the initiation of rapidly increasing power energy demand, at the present it become necessary to combine a non-conventional generating systems like

WES to support the existing conventional generating system. The outcome of incorporating WES on conventional ED problem makes the optimization problem complex, non-linear and non-convex due to stochastic nature of wind.

3.3.1 Problem Formulation

The objective of the EDWES problem is to minimize the total fuel costs of available conventional generating units with WES subject to equality and inequality constraints. The mathematical model can be defined as (Sanjay Roy 2012).

$$\min f = \sum_{j=1}^{ng} Fc_j \left(P_{tp,j} \right) + \sum_{m=1}^{nw} F_{WES,m} \left(P_{WES,m} \right) \tag{3.8}$$

3.3.2 Wind Energy System

The total generation cost of the WES according to (Sanjay Roy 2012) the quadratic cost function of mth WES units is expressed as,

$$F_{WES,m} \left(P_{WES,m} \right) = wc_m P_{WES,m} \tag{3.9}$$

The expression for electrical power output of mth WES involving turbulence (Rozenn et al. 2010), (Zhang et al. 2008) as at the hub of the wind turbine, speed variations were observed at the blades linked to the central shafts is given in Equation (3.10). This expression is obtained by neglecting the impact of long duration variation of wind speed at WES installation sites and with no proportionality of stable mean wind speed, u_m^{mw} with WES power output. In a pitch angle controlled WES, the power output from WES is given by the relation between $P^{rat}_{WES,m}$ and $\mu_{weq} \left(u_m^{mw}, l- \right)$ using an analytical expression of standard output as

$$P_{WES,m} = \mu_{weq}\left(u_m^{mw},1\right)P_{WES,m}^{rat} \tag{3.10}$$

As suggested in Rozenn et al. (2010), Manwell et al. (2009), impact of long period variation of wind speed is ignored and only the short period wind speed turbulence intensity r is considered in Equation (3.21). The appropriate expression of the per unit output under turbulence $\mu_{weq}\left(u_m^{mw},1\right)$ from experimental data is used in ED problems (Sanjay Roy 2012). Hence, the per unit output of WES considering turbulence intensity r is expressed as,

$$\mu_{weq\,m}\left(u^{mw},1\right)=1-\exp\left(-\frac{uw_m^{mw}/uw_m^{m\,rat}}{vw(1)^{kw(1)}}\lambda\,I\right) \tag{3.11}$$

$vw(1)$ and $kw(1)$ which can be described as (Benhamida et al. 2013),

$$kw(1)=3.49-6.14\,1 \tag{3.12}$$

$$vw(1)=0.71-0.21\,1-1.26\,1^2 \tag{3.13}$$

Equation (3.12) and (3.13) are utilized for estimating the output under turbulence $\mu_{weq}\left(u_m^{mw},1\right)$. Once rated output of WES is known, the power generated from WES can be calculated.

3.3.3 Equality Constraint

The equality constraint ensures that the total power is balanced between generation, demand and losses, where the total real power generations from thermal generating units plus WES meet the required power demand plus transmission losses of electrical network (Sanjay Roy 2012).

$$\sum_{j=1}^{ng}P_{tp,j}+\sum_{m=1}^{nw}P_{WES,m}=P_D+Pt_{Loss} \tag{3.14}$$

3.3.4 Thermal Bounded Power Limits

Thermal generating inequality constraints of EDWES are considered in section 3.2 .

3.3.5 WES Power Limits

The real power out of each WES should be within the maximum generating limits of rated WES, which is given below

$$0 \succ P_{WESm} \quad \succ P_{WES,m}^{rat} \tag{3.15}$$

3.4 CAPTURE OF MAXIMUM POWER FROM WIND ENERGY SYSTEM

The wind power system design must maximize the annual energy capture at a given site. The maximum power capture algorithms researched so far can be classified into three main control methods, namely Tip-Speed Ratio (TSR) control, Power Signal Feedback (PSF) control and Hill-Climb Search (HCS) control. The most common operating mode for extracting the maximum energy (Mukund 2006) is TSR control. TSR control is to vary the turbine speed along with varying wind speed such way that at all the times the TSR is continuously equal to that required maximum power co-efficient C_{pw} .

In theoretical and field experience indicates that the variable-speed operation yields 20 to 30% more power than with the fixed-speed operation. However, the cost involved in variable-speed control is added for generation of electrical power.

In the typical system design, the trade-off between energy increase and cost increase has to be optimized. In recent years, the inclusion of cost of designing the variable pitch rotor and speed control with power electronics overshadows the benefit of the increased energy capture. However, the falling prices of power electronics for speed control and the availability of high-strength fibre composites for constructing high-speed rotors have made it economical to capture more energy when the speed is high.

The variable-speed operation has an indirect advantage. It allows controlling the active and reactive powers separately in the process of automatic generation control. In fixed-speed operation, on the other hand, the rotor is shut off during high wind speeds, losing significant energy.

Therefore, operating the wind turbine at a constant TSR corresponding to the maximum power point at all times can generate 20 to 30% more electricity per year (Mukund 2006). Hence, this requires a control scheme to operate with a variable speed to continuously generate the maximum power.

This section, maximizing wind energy capture make use of optimizing TSR in most common Constant Frequency (CF) scheme as an substitute of using Variable Frequency (VF) scheme, in view of the actuality that VF has the major disadvantage in controlling independently the speed of each turbine.

The organization of this section 3.5 formulates generation of wind energy from typical wind turbine for constant frequency scheme. The control parameters associated with generating wind energy such as TSR, power co-efficient are explained in this section. Section 3.6 formulates optimization of Capture of Maximum Wind Energy (CMWE) problem for constant frequency scheme. Also in this section, briefing the generation of power from the wind turbine to minimum value of power co-efficient by TSR.

3.5 PROBLEM STATEMENT

3.5.1 Wind Turbine Power Generation

The generated power output of a typical wind turbine is a function of wind speed cube and power co-efficient which is depending on the specific wind characteristic and optimal power co-efficient C_{pw} (Oriol Gomis-Bellmunt et al. 2010),

$$Pg_w = \frac{1}{2}\pi A v_w^3 C_{pw}\left(l\varpi_{TSR}, \lambda_{pitch}\right) \tag{3.16}$$

3.5.2 Power Co-Efficient

The fraction of the upstream wind power that is extracted by the rotor blades and fed to the electrical generator (Mukund 2006) the remaining power is dissipated in the downstream wind as power co-efficient or rotor efficiency, C_{pw}. The maximum rotor efficiency C_{pw} is achieved at a particular value TSR.

The maximum power is extracted from the wind at that speed ratio, i.e., when the downstream wind speed equals one third of the upstream speed. Under this condition theoretical maximum value of C_{pw} is 0.59 (Mukund 2006).

Under the theoretical study the maximum value of C_{pw} is at 0.59. C_{pw} can often expressed as a function of TSR as shown in Figure 3.6. Thus TSR is linear speed of the rotors outermost tip to the upstream wind speed. The relationship between the rotor tip speed and the wind speed is established with the aerodynamic analysis of the wind flow around the moving blade with a given pitch angle. In practical designs, the achievable maximum rotor efficiency C_{pw} varies between 0.4 and 0.5 for modern high speed two-blade turbines, and for slow-speed turbines between 0.2 and 0.4 with more blades. If C_{pw} value is considered as 0.5, then the rotor efficiency is practically maximum (Mukund 2006).

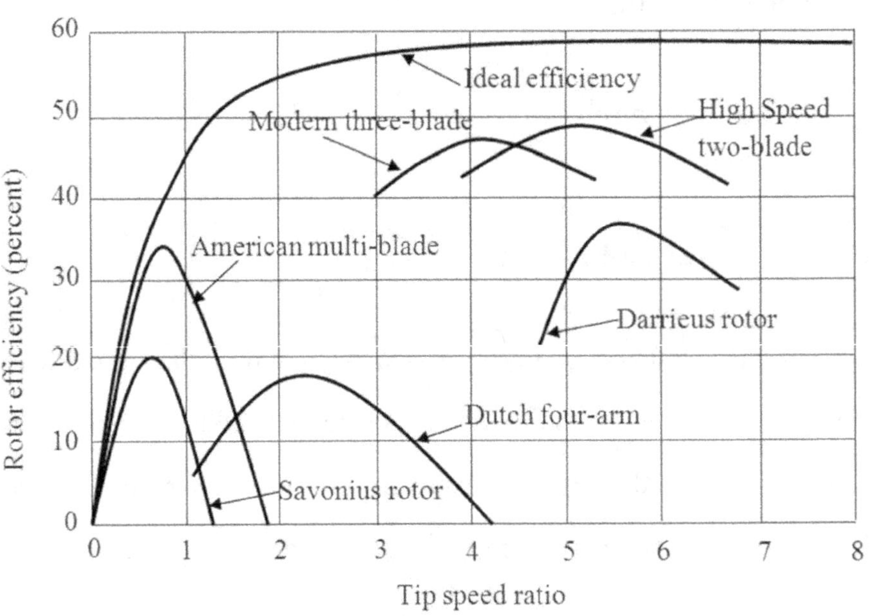

Figure 3.6 Rotor efficiency vs TSR with different number of blades

In general, the power co-efficient, C_{pw} is often expressed as a function of the rotor Tip-Speed Ratio (TSR) as (Oriol Gomis-Bellmunt et al. 2010),

$$C_{pw}\left(\lambda_{TSR},\beta_{pitch}\right) = c_1 K_{cp} e^{-c_6 \frac{1}{\lambda_{TSR}}} \tag{3.17}$$

where,

$$K_{cp} = c_2 \frac{1}{\lambda_{TSR}} - c_3 \beta_{pitch} - c_4 \beta_{pitch} - c_5 \beta_{pitch} - c_6 \tag{3.17a}$$

3.5.3 Tip-Speed Ratio

TSR is the linear speed of the rotor's outermost tip to the upstream wind speed. The aerodynamic analysis of the wind flow around the moving blade with a given pitch angle establishes the relation between the rotor tip-speed and the wind speed (Mukund 2006) is given by,

$$\lambda_{TSR} = \frac{\mu R}{v_w} \tag{3.18}$$

3.5.4 Typical Maximum Operating Point of Wind Turbine

The typical maximum operating point of the wind turbine is when the maximum value of C_{pw} is at 0.59 is called as Beltz limit. The maximum operating point of can be achieved for the wind speed should not violate the maximum threshold value C_{pw}. Therefore, the maximum C_{pw} of SWT (Oriol Gomis-Bellmunt et al. 2010) is given by

$$C_{pw}^{max}\left(\lambda_{TSR},C_{pw}^{max}\right) = c_1\left(c_2 \frac{1}{12} Ie^{-\frac{c_6 c_7}{c_2}-1} \therefore c_7\right) \tag{3.19}$$

Typical TSR - C_{pw} characteristics of SWT is shown in Figure 3.7.

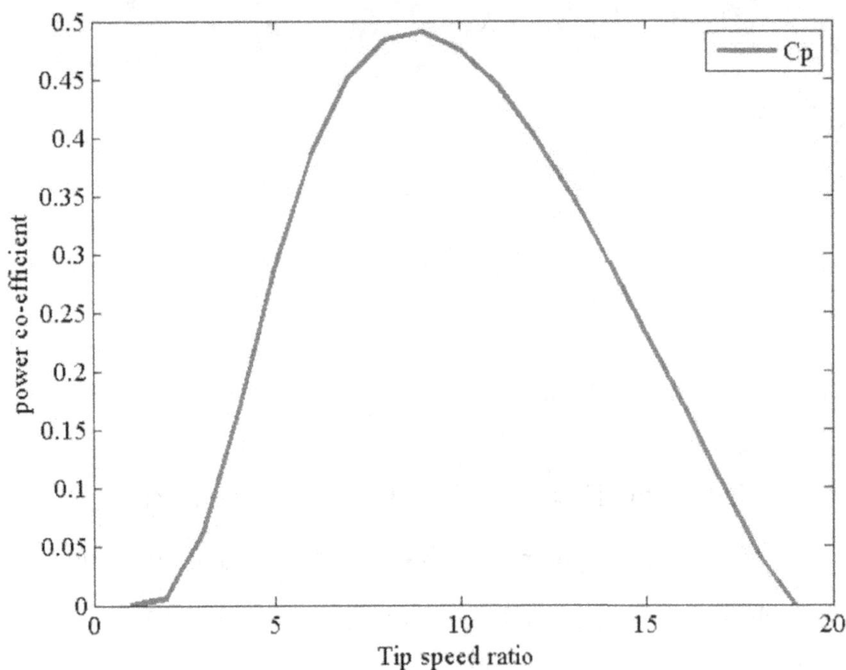

Figure 3.7 Typical TSR - C_{pw} characteristics of SWT

3.6   PROBLEM FORMULATION OF CMWE

Without loss of generosity, wind turbine in the off shore wind farm of constant frequency can be expressed as according to (Oriol Gomis-Bellmunt et al. 2010).

$$\max\left(Pg_w\right) = \frac{1}{2}\,\pi A v^3\, C_{w\,pw}^{optimal}\left(A_{TSR}, \lambda_{pitch}\right) \tag{3.20}$$

$$C_{pw}^{optimal}\left(A_{TSR}\right) = 0.44\left(K\right)e^{-165\left(\frac{1}{\therefore A_{TSR}}+0.0021\right)} \tag{3.21}$$

where,

$$K = 125\left(\frac{1}{\therefore A_{TSR}}+0.0021-6.94\right) \tag{3.21a}$$

According to (Oriol Gomis-Bellmunt et al. 2010) the power co-efficient $C_{pw}^{optimal}$ can be evaluated as the pitch angle is assumed to 0, C_{pi} is computed using the Equation (3.21) is the power co-efficient of a typical single wind turbine is considered.

3.7 CONCLUSION

The following types of the problems are formulated in this section: 1) ED of conventional thermal power generators with having transmission loss bounded generating limits, prohibited zones and ramp-rate limits, 2) EDWES problem of incorporating WES with turbulence to conventional ED problem are described. 3) problem formation for generation of electrical power from a typical SWT and 4) without loss of generosity, formulating the CMWE problem based on considering the power co-efficient formulation of a typical referred 2MW rated wind turbine. In the following Chapter 5, section 5.5, the proposed COA is presented to solve the CMWE problem.

CHAPTER 4

MODELING OF BIO-INSPIRED CAPRA OPTIMIZATION ALGORITHM

4.1　　INTRODUCTION

The conventional ED, EDWES and CMWE problems are real world optimization involving real numbers as their parameters. Many local optima are achieved in the objective function due to these parameters which lead to highly non-linear. Hence, developing an optimization technique will be useful to solve the problem with these highly non-linear characteristics. In recent past, optimization techniques such as GA, PSO, DE, and so on has been demonstrated effectively for solving problems in which the objective function is considered as high dimensional; with reliable-variable and having multiple local optima.

The success of these well known algorithms largely depends on the coding of the problem variables and on the crossover operators. However lot of difficulties needed to be faced in the coding of real-valued variables in achieving precision in the solution obtained. In order to overcome the above difficulties, many attempts have been made in recent times. Thus, there is a need to solve optimization problems having high non-linearity in a continuous search space. Therefore, this section proposes a new optimization approach called Capra Optimization Algorithm (COA) to solve ED, EDWES and CMWE.

In this thesis, this chapter proposes the newer optimization technique called COA along with theoretical development of COA. The efficiency and effectiveness of COA algorithm is applied to ED, EDWES and CMWE by comparing the results obtained from other well-known optimization algorithms reported in the literatures are described in chapter 5.

The development of COA is presented below:

4.2 THEORITICAL DEVELOPMENT OF COA

A COA is developed from bio- inspired behaviour of an herbivore genus known as Capra (domestic goat). Capra is an herbivore animal anatomically adapted to eating (feeding) plant materials for their diet. Herbivora is derived from Latin word "Herba" meaning a small plant and "vora" means to devour (eat) (Glare 1990). Feeding strategies of all herbivores employ either browsing or grazing.

Browsing is a kind of feeding type in which herbivore consumes leaves of soft shoots, fruits of high-growing plants (Nicholson 1970). Browsing has its own disadvantages when it is to be considered for the development of an optimization algorithm. Since there is a lack of constant permanent supply of food which results in discontinuous searching capability draws a conclusion that the browsing type herbivore is not suitable for development of an optimization algorithm.

Grazing is another type of feeding technique in which herbivore consumes growing grass and pasture i.e., grazing grass field (Helen Armstrong et al. 1993). Since the availability of food is continuous which results in continuous searching capability, grazing can be considered in the development of an optimization algorithm. In addition, herbivore which follows grazing type of feeding technique doesn't stores the food and consumes it without skipping.

Thus the above stated difference in the herbivore feeding techniques results in selecting the grazing type feeding technique for developing an optimization algorithm in this thesis.

4.2.1   Reason in Selecting Capra

Capra has exceptional behaviour which makes it different from other grazing type herbivore species. In general, the grazing herbivore species is selected for an application based on the parameters like selective ability, minimum sward height and biting methods. Selective ability refers to the ability of herbivore to select the proper food. Sward height refers how much size an herbivore can intake at single bite (Helen Armstrong 1993). Bite is a common behaviour which involves opening and closing of the jaw found in herbivores. Some examples of grazing species are Sheep, Red deer, Cattle, Capra, Horses and Rabbits (Helen Armstrong et al. 1993).

Among these species the grazing behaviour of Sheep is highly selective but has only 3 cm minimum sward height grazing. Red deer has selective ability is medium and minimum sward height is around 4 cm. Cattle has selective ability is slightly selective, minimum sward height is greater than 6 cm. Horse has minimum selective range and minimum sward height is

2 cm. Rabbit has very high selective ability but the minimum sward height is only 1 cm. Capra has highly selective ability and also have the minimum sward height is greater than 6 cm (Appendix 1).

Thus, Capra is more capable of utilizing natural grazing land since it has highly selective ability which covers wide area in search of feeding. Also Capra finds the most nutritious available feed from the grazing land in reasonable time. This motivates the researchers to model the behaviour of Capra into an optimization technique which can be called as COA.

4.3　MODELING OF COA

In modeling of COA, the Capra behaviour of utilizing its selective ability in a grazing land (for exploration) is considered as search space. In addition, the ability of migrating (for exploitation) towards the nutritious food and eating the food with the bite rate (for increasing exploitation speed) in the typical grazing land is considered as the ability of the proposed COA algorithm to reach (obtain) the global optimal solution.

The different parameter in the proposed COA which helps in exploring and exploiting the promising regions of search space are:
1) Grazing land type, 2) Length of the rope where Capra is tied and 3) bite rate, which are discussed below.

4.3.1　Grazing Land

The exploration region of the search space in real world optimization problem is considered as grazing land in the proposed COA. The different grazing land which is denoted as (ξ) where Capra can grazes are:

mountains, grassland, health land, meadows, rough pasture and rangeland (Elfyn et al. 1958). But for solving the numerical optimization problem, the grazing land type is considered unit circle i.e., $\xi = 1$. In this thesis, the area grazed by the Capra is equal to half the area of the ξ for exploiting the space in the search region.

4.3.2   Length of the Rope

The success of COA in obtaining the global solution depends on the grazing of Capra which is determined by the length of the rope where the Capra is tied. Hence, it can be concluded that the reachable exploitation of Capra is mainly depends on the length of the rope lr which is determined as follows:

Consider a fenced circular grazing land ξ with known radius R . At the edge of grazing land type consider a pole with a rope attached to it. At the other end of the rope a Capra has been tied. The above situation is described in the Figure 4.1.

Figure 4.1 shows circular grazing land type ξ of radius R centered at O, and the rope is attached to the pole present at the fence at point B . The limit of the Capra's tether is the circle of radius r through C centered at B . The upper half of the section accessible by the Capra consists of a portion of the circle of radius r subtended by the angle δ , plus a portion of a unit circle subtended by the angle $<$, minus the triangular region OCB.

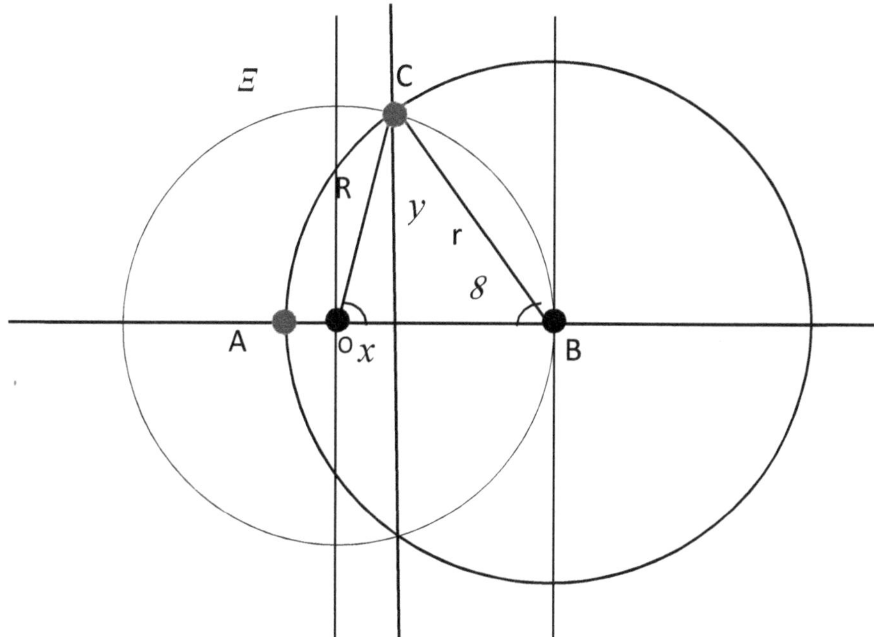

Figure 4.1Grazing area of Capra

According to (Kmath 2014) typical reachable area of Capra is equal to some specified fraction $\varphi\lambda$ (such as one half) of the area of the upper half of the grazing land type $\varXi$. Therefore, the area grazed by the Capra is given by

$$\frac{\delta}{2v}\left(\chi v r^2\right)+\frac{}{2v}\left(\chi v R^2\right)-\frac{y}{2}=\frac{\varphi v}{2} \tag{4.1}$$

For an optimization problem, the grazing land type is considered as an unit circle i.e., $R=1$. Substituting $R=1$ in the above equation, we get

$$\frac{\delta}{2v}\left(\chi v r^2\right)+\frac{}{2v}\left(\chi v\right)-\frac{y}{2}=\frac{\varphi v}{2} \tag{4.2}$$

Simplifying the above equation, we get,

$$\delta r^2+-y=\varphi v \tag{4.3}$$

The following assumption is made to solve the above equation: *CB* is an arc of the circle centred at O such that the angle OCB equals the angle OBC and hence

$$\delta = (\nu - \beta)/2.\tag{4.4}$$

Using the right angled triangles ODC and CDB, we get

$$r^2 = (1-x)^2 + y^2 \text{ and}$$

$$y^2 = 1 - x^2$$

Substituting y^2 in the above equation and simplifying, we get

$$r^2 = 2(1-x)\tag{4.5}$$

Using Equation (4.4) and Equation (4.5) in Equation (4.2), we get,

$$2(1-x)\frac{\nu - \beta}{2} + \beta - y = \pi + \nu\tag{4.6}$$

Also, from triangle ODC, we get

$$x = \cos\beta \text{ and } y = \sin\beta$$

Substituting the above relation in equation (4.6), we get

$$(1-\cos\beta)(\nu - \beta) + \beta - \sin\beta = \pi + \nu\tag{4.7}$$

Using the relations $\alpha = \nu - \beta$ and $\beta = \nu - \alpha$ in the above equation, we get

$$(1-\cos\alpha)\alpha + \nu - \alpha - \sin(\nu - \alpha) = \pi + \nu\tag{4.8}$$

Since, $\gamma = v - \alpha$, we get

$$\sin\gamma = \sin(v-\alpha) = \sin\alpha \text{ and}$$

$$-\cos\gamma = \cos(v-\gamma) = \cos\alpha$$

Substituting the above relations in equation (4.8), we get

$$(1+\cos\alpha)\alpha + v - \alpha - \sin\alpha = \mu v \tag{4.9}$$

Simplifying the above equation, we get

$$\sin\alpha - \alpha\cos\alpha = (1-\mu)v \tag{4.10}$$

By evaluating Equation 4.10 α is been determined for any fraction of μ . The exploitation of Capra in the search space by the length of the rope is determined as (Kmath 2014) ,

$$r = \sqrt{2(1+\cos\alpha)} \tag{4.11}$$

The Equation 4.11 finds the area grazed (reachable exploitation) by Capra in the unit circle which is the length of the rope r and is used to generate the different population in COA

4.3.3   Bite Rate

After reachable area of Capra, bite count creates a new vector corresponding to each population member of COA through bite rate. In this study Capra has been restricted to 100-150 nominal bites (Gong et al. 1996) with the intention of diminishing overlapping bites and the time between the

first and the last bite has been recorded from bite number, bite rate and bite
time.

$$f_{rate} = \frac{f_{count}}{T_h * f_{strength} * f_{sward}} \tag{4.12}$$

4.4 GENERATING INITIAL POPULATION

Initialize a population $x_{i,j}$ individuals with random values
generated of N real-valued parameter vectors. The real value parameter is
accurate and efficient because it is close to the real design search space. Here
a vector $x_{i,1}, x_{i,2}, ..., x_{i,N}$ is a ith parent vector to represent a solution to an
optimization problem. Initially, M number of parent vectors is generated
using,

$$x_{i,j} = x_j^{max} \xi rand(0, lr) \tag{4.13}$$

where $i = 1, 2, ..., M$ $j = 1, 2, ..., N$, $rand(0, lr)$ is a uniformly distributed random
number lying between 0 and lr. If a parent vector $x_{i,j}$ violates the operating
limit constraints, then it is made to satisfy the operating constraints and then
fitness function $f(x_i)$ is calculated. In the COA the M number of parent
vectors are ranked against each other based on the objective function value
and the one with minimum value is the best position of the Capra and it is
taken as $x_{best,j}^t$

4.5 GENERATING NEW POPULATION

After initialization, COA generates a vector $z^t_{i,j}$ using mutation process which is based on bite rate. When the creation of offspring is closer to the parents as the generation t proceeds, the probability of creating offspring closer to the parents will be higher and higher, and the offspring is created in the following equation

$$z^t_{i,j} = x^t_{best,j} + rand\left(0, \lambda_{rate} * lr\right) * \left(x^t_{R1,j} - x^t_{R2,j}\right) \qquad (4.14)$$

where, R_1 and R_2 are mutually different integers that are also different from the running index i. COA uses, the probability of $\left(\lambda * lr_{rate}\right)$ as a positive control parameter for scaling the difference between vectors of individuals with indexes R_1 and R_2 to avoid stagnation.

4.6 BINOMIAL RECOMBINATION

Following the generation of newer population, COA uses the binomial recombination process. Binomial recombination is employed to generate a vector based on the relation between the randomly generated number between 0 and lr and cross over value CRX. The CRX recombination rate lies in the range of [0, 1]. The binomial recombination employed as in the following equation as,

$$u^{t+1}_{i,j} = \begin{cases} z^t_{i,j} & \text{if}\left(rand[0,lr] \geq CRX\right) \\ x^t_{i,j} & \text{otherwise} \end{cases} \qquad (4.15)$$

where $rand\,[0,lr]$ is uniformly distributed random numbers

4.7 SELECTION PROCESS

In the selection process, the existence of offspring generated in the recombination process U_i^{t+1} in the population for the next generation is decided. In selection process, U_i^{t+1} is compared with the corresponding parent vector x^t based on the relation given below:

$$X^{it+1} = \begin{cases} U_i^{t+1} & \text{if } f\left(U_i^{t+1}\right) : \sigma\, f\left(X_i^t\right) \\ X^t & \text{otherwise} \\ \lambda_i \end{cases} \tag{4.16}$$

4.8 STOPPING CRITERION

COA technique stops when the specified maximum number of generations (Iterations) $t_{\max}$ is reached.

Thus, COA involves two important issues: 1) progressive search space direction and 2) population multiplicity. A progressive strong search space direction is effective in searching as well as to reduce the computational burden and to increase the probability of finding an optimal solution quickly. Population multiplicity creates the new generation of offspring that differs more from parents in the search space. In view of these highly diverse newly generated population can prevent the premature convergence to a local optimum and also increases the probability of exploiting the global optimum solution.

4.9 IMPLEMENTATION OF COA

The steps involved in COA are represented in Figure 4.2 and are described as follows:

Step 1: Initialization of parameters: population size M, number of variables N, binomial crossover CRX, grazing land type Ξ, bite count γ_{count}, bite time T_{jt}, bite strength $\gamma_{strength}$, sward height, γ_{sward} maximum number of generation t_{max}

Step 2: Evaluation of Capra parameters: length of the rope lr, bite rate γ_{rate} are calculated using Equations (4.5) and (4.6) respectively.

Step 3: Generating initial population: generating the initial population $x_{i,j}$ using Equation (4.13) and then the fitness function is calculated for the generated initial population also determine the best individual x^{best}_{j} of the initial generated population.

Step 4: Generation of new population: Generating new set of population, $z^{t}_{i,j}$ with the help of bite rate of the Capra using the Equation (4.14).

Step 5: Binomial recombination: Binomial recombination $U^{t+1}_{i,j}$ using *CRX* operator as in Equation (4.15).

Step 6: Selection process: select the best individual $x^{t+1}_{i,j}$ using Equation (4.16).

Step 7: Check for stopping criterion, if the stopping criterion t_{max} is met go to step 8 Otherwise go to step 3.

Step 8: Obtain the optimal solution and the optimal fuel cost.

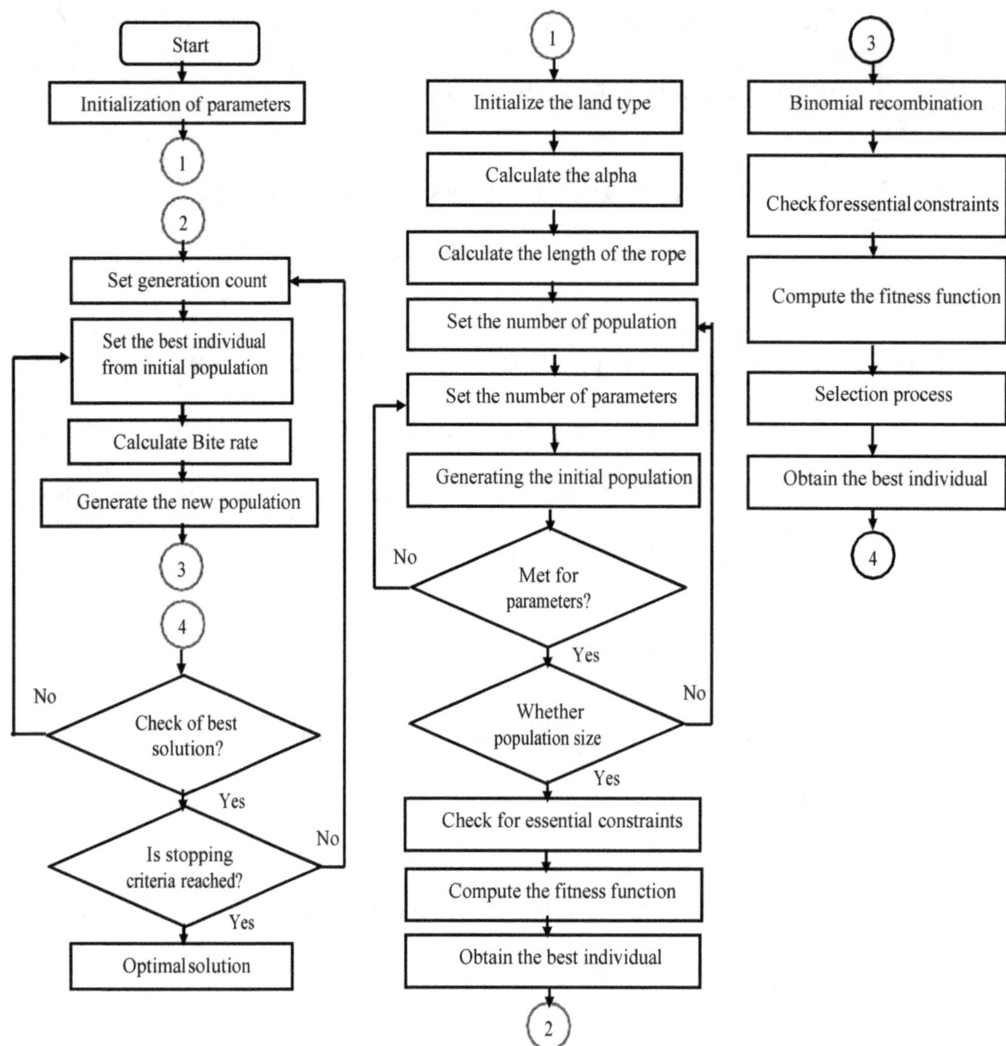

Figure 4.2 Flowchart of Capra optimization algorithm

4.10   CONCLUSION

In this chapter, a new optimization technique called COA along with the theoretical development of COA has been described. The efficiency and effectiveness of COA algorithm is applied to ED, EDWES and CMWE by comparing the results obtained from other well-known optimization algorithms reported in the literatures are described in chapter 5.

CHAPTER 5

RESULT AND DISCUSSION

5.1   INTRODUCTION

In this chapter, the proposed COA is successfully applied to the ED, EDWES and CMWE problems and the results obtained are compared with various optimization methods. The organizations of this chapter are section 5.2 discusses the various test systems implemented on the proposed COA algorithm to ED, EDWES and CMWE problems. Section 5.3 discusses the determination of COA parameter. Section 5.4 discuses the detailed simulation results of proposed COA in solving conventional ED problems applied to the test system-1, test system-2, test system-3 and test system-4. Section 5.5 deals with the simulation results of proposed COA in solving conventional ED problem incorporating single WES to existing test system-1, test system-2, test system-3 and test system-4. In section 5.6, is briefed the detailed simulation results of proposed COA in solving CMWE for constant frequency off shore wind farm.

5.2   TEST   SYSTEM   DESCRIPTION   USED   IN   THE SIMULATION

The efficiency of the proposed COA is illustrated solving the ED, EDWES and CMWE problems by considering the test systems taken from literature. The following are the detailed description of test systems which are considered and the simulation results are discussed in chapter 5 of this thesis.

5.2.1 Test System-1: Conventional Thermal Power Plant Having 3-Generating Units

The data for the test system-1 comprising of 3-generating units of conventional thermal power plant are taken from (Po-Hung et al. 1995). The thermal generator cost co-efficient and B-Loss co-efficient given in Appendix 2. The total power demand of the system is 300MW

5.2.2 Test System-2: Conventional Thermal Power Plant having 6-Generating Units

The data for the test system-2 comprising of 6-generating units of conventional thermal power plant with ramp-rate and prohibited operating zones are taken from (Gaing 2003). The thermal generator cost co-efficient and B-Loss co-efficient is given in Appendix 3. The total power demand to meet up is 1263 MW.

5.2.3 Test System-3: Conventional Thermal Power Plant having 15-Generating Units

The data for the test system-3 comprising of 15-generating units of conventional thermal power plant with prohibited operating zones in units 2, 5, 6 and 12 are taken from (Gaing 2003). The thermal generator cost co-efficient and B-Loss coefficient is given in Appendix 4. The power demand to meet up is 2630 MW.

5.2.4 Test System-4: Conventional Thermal Power Plant having 20-Generating Units

The data for the test system-4 comprising of 20-generating units of conventional thermal power plant is taken from (Su & Lin 2000). The thermal generator cost co-efficient and B-Loss co-efficient is given in Appendix 5 . The power demand to meet up is 2500 MW.

5.2.5   Test System-5: Test system-1 with WES

According to (Sanjay Roy 2012) single WES unit has been considered for this test system in addition to 3   conventional   thermal generating units existing in test system-1. The rated   output   of   WES considered for including WES to test system-1is 100 MW which is assigned based on the assumption presented in (Sanjay Roy 2012). The cost   co-efficient wc_m of the WES unit has been calculated from the conventional thermal unit cost co-efficient 37.55% of $b_{tp,3}$ of test System-1

5.2.6   Test System-6: Test System-2 with WES

According to (Sanjay Roy 2012) single WES unit has been considered for this test system in addition to 6   conventional   thermal generating units existing in test system-2. The rated   output   of   WES considered for including WES to test system-2 is 500 MW which is assigned based on the assumption presented in (Sanjay Roy 2012). The cost   co-efficient wc_m of the WES unit has been calculated from the conventional thermal unit cost co-efficient 37.55% of $b_{tp,1}$ of test system-2

5.2.7   Test System-7: Test System-3 with WES

According to (Sanjay Roy 2012) single WES unit has been considered for this test system in addition to 15 conventional thermal generating units existing in test system-3. The rated   output   of   WES considered for including WES to test system-1is 455 MW which is assigned based on the assumption presented in (Sanjay Roy 2012). The cost co-efficient wc_m of the WES unit has been calculated from the conventional thermal unit cost co-efficient 37.55% of $b_{tp,1}$ of test system-3.

5.2.8   Test System-8: Test System-4 with WES

According to (Sanjay Roy 2012) single WES unit has been considered for this test system in addition to 20 conventional thermal generating units existing in test system-4. The rated output of WES considered for including WES to test system-1is 600 MW which is assigned based on the assumption presented in (Sanjay Roy 2012). The cost co-efficient wc_m of the WES unit has been calculated from the conventional thermal unit cost co-efficient 37.55% of $b_{tp,1}$ of test system-4. WES is incorporated as the short time duration wind speed variation effect as a static power and not as stochastic models for the simulation of test system-5 to test system-8.

5.2.9   Test System-9: Wind Energy System for CMWE

The data for the test system-9 comprising of single generating units of WES for constant frequency off shore wind farm is taken from (Oriol Gomis-Bellmunt et al. 2010). The rated electrical power output of WES is 2MW and detailed description of WES is represented in Appendix 6.

COA simulation has been carried out for ED, EDWES and CMWE are coded in Matlab 7.0 version language, executed on Pentium P4, Core 2 duo processor, 2.4 GHz and 1 GB RAM personal computer. The performance of COA is studied by running the program for 100 trail runs. The optimal fuel cost obtained in each test systems is used to compare the performance of COA with other methods. The dollar ($) is converted into Indian rupee (`)

with conversion value as on 19-05-2017 (\$ 1= ` 64.46) for the comparison in Indian rupees.

5.3   DETERMINATION OF COA PARAMETER

The control parameter grazing area $\because\varphi\vartheta$ in COA which influences quicker the optimal solution is determined and the same value of grazing area $\because\varphi\vartheta$ is used for entire simulations for test systems in section 5.2. For determining this grazing area $\because\varphi\vartheta$ simulation is carried out for 100 trail runs of COA with the selected population size of 25. The minimum total generation cost is obtained by solving conventional ED problem for different values of the grazing area $\because\varphi\vartheta$ applied to test system-1, test system-2, test system-3 and test system-4 is shown in Table 5.1.

Table 5.1 Determination of COA parameters

Grazing area $\because\varphi\vartheta$	Length of the rope lr	Grazing angle α	Minimum total generation cost in ` /h			
			Test System-1	Test System-2	Test System-3	Test System-4
0.10	0.4746	2.6625	2352504	995791	2144693	4033922
0.15	0.5863	2.5466	2352483	995787	2144604	4033846
0.20	0.6849	2.4425	2352419	995753	2144526	4033847
0.25	0.7748	2.3460	2352419	995749	2144483	4033845
0.30	0.8584	2.2544	2352419	995745	2144244	4033851
0.35	**0.9376**	**2.1658**	**2352419**	**995716**	**2144228**	**4033842**
0.40	1.0135	2.0788	2352479	997068	2144745	4033953

From Table 5.1, it is observed that, the best parameter value of COA is of grazing area $\because\varphi\vartheta$ at 0.35 is used for entire simulations for test systems, length of the rope $lr = 0.9376$ and the grazing angle $\alpha = 2.1658$.

Therefore, in this thesis, the COA parameters obtained using table 5.1 is used for implementing proposed COA technique to conventional ED, EDWES and CMWE problems.

5.4 APPLICATION OF COA TO CONVENTIONAL ED PROBLEM

The performance of COA is evaluated by implementing to solve conventional ED problems. The detailed description of each test systems is given in section 5.2. To verify the performance of COA, the program is run for 100 trail runs for the population size $M = 100$. The minimum fuel cost, maximum cost, mean cost, standard deviation and the execution times are used to compare the performance of COA with other methods.

5.4.1 Test System-1

The test system-1 is a conventional 3-generator unit with prohibited operating zones with a total demand of 300MW. The results obtained for the test system-1 by implementing COA is tabulated in Table 5.3. The comparison of minimum total generation cost for Test System-1 is tabulated in Table 5.2

From Table 5.2, it is observed that the minimum total generation cost obtained using the proposed COA is ` 232650.32 which results in a saving of ` 8249.59, ` 2796.28, ` 1617.30, ` 1191.87, ` 1126.76, ` 706.48 , ` 702.62, ` 679.41, ` 502.14 and , ` 278.47 when compared with GA, T-NN, APSO, SARGA, LAM-ITR, BBO, BGA, DE/BBO, MPSO and COA respectively

Table 5.2 Comparison of minimum total generation cost for test system-1

Optimization methods	Minimum total generation cost	
	$	`
GA (Naresh & Dubey 2004)	3737.20	240899.91
T-NN (Naresh & Dubey 2004)	3652.60	235446.60
APSO (Muthu & Thanushkodi 2012)	3634.31	234267.62
SARGA (Subbaraj et al. 2010)	3627.71	233842.19
Lambda iteration (Naresh & Dubey 2004)	3626.70	233777.08
BBO (Muthu & Thanushkodi 2012)	3620.18	233356.80
BGA	3620.12	233352.94
DE/BBO (Muthu & Thanushkodi 2012)	3619.76	233329.73
MPSO	3617.01	233152.46
SADE	3613.54	232928.79
COA	3609.22	232650.32

Electrical power output generation of each generator and minimum total generation cost of various methods are listed in Table 5.2. From the table 5.2, it may be noted that the performance of COA is better than other algorithms in terms of minimum total generation cost and execution time.

For the test system-1, statistical performance is analyzed by conducting 100 independent trail runs for the given power demand. The results of minimum mean and maximum total generation cost, standard deviation and computational time are tabulated in Table 5.4.

Table 5.3   **Generation of electrical power output (MW) and minimum total generation cost for test system-1**

Unit power Output (MW)		APSO (Muthu & Thanushkodi 2012)	SARGA (Subbaraj et al. 2010)	BBO (Muthu & Thanushkodi 2012)	MPSO	SADE	COA
P_1		200.53	192.90	207.99	219.35	211.38	214.61
P_2		78.28	85.01	86.01	68.32	77.46	78.63
P_3		34.00	34.08	16.07	22.80	21.13	15.93
Transmission Loss (MW)		12.79	11.99	10.07	10.47	9.97	9.17
Total Electrical Power Output (MW)		312.79	311.99	310.07	310.47	309.97	309.17
Minimum total generation cost ($/h)	$	3634.31	3620.17	3619.76	3617.01	3613.54	3609.26
	`	234267.62	233842.19	233329.73	233152.46	232928.79	232650.32

The convergence characteristic of COA for test system-1 is shown in Figure 5.1 along with well known methods BGA, MPSO and SADE. Convergence characteristics have been plotted for 100 number of generations (Iteration) to total generation cost in $. It is inferred from the graph that COA has superior features including high quality solutions in terms of total minimum generation cost, stable convergence characteristics and good computational efficiency.

Table 5.4 Statistical analysis of various methods for test system-1

Optimization Methods	Minimum total generation cost in $	Mean total generation cost in $	Maximum total generation cost in $	Standard deviation	Execution time in seconds
BGA	3620.12	3620.23	3622.87	0.3365	4.23
MPSO	3617.01	3617.17	3619.75	0.3446	4.28
SADE	3613.54	3613.55	3613.35	0.6475	4.71
COA	3609.22	3609.29	3611.39	0.2943	1.58

The effectiveness of COA over the other well known optimization methods is established by attaining a total generation cost within the specific ranges (out of 100 trail runs). To compare this, the results among BGA, MPSO, SADE and COA in a statistical manner, relative frequency of convergence in the range of total generating cost for test system-1 is tabulated in Table 5.5. From Table 5.5, it is observed that the successfulness of COA over other well known optimization methods is established in terms of high quality solution and better convergence.

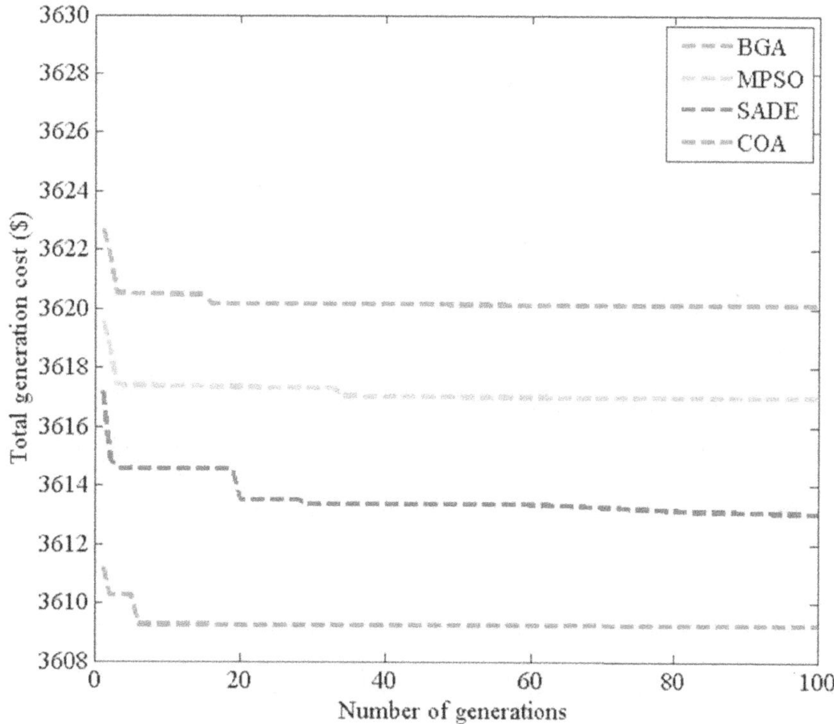

Figure 5.1 Convergence characteristics for test system-1

Table 5.5 Relative frequency of convergence in the range of total generation cost of various methods for test system-1

Range of total generation cost in $	3622.86-3620.16	3620.16-3617.39	3617.39-3617.34	3617.34-3164.59	3614.59-3613.32	3613.32-3610.28	3612.28-3610.26	3610.26-3609.23	3609.23-3609.23
BGA	15	85	-	-	-	-	-	-	-
MPSO	-	8	92	-	-	-	-	-	-
SADE	-	-	-	6	57	37	-	-	-
COA	-	-	-	-	-	2	3	10	85

5.4.2   Test System-2

This test sytem-2 consists of 6-generating units with prohibited operating zones. The total demand for test system is 1263MW. The result obtained by proposed COA is tabulated and compared in Table 5.6, Table 5.7, and Table 5.8 respectively. From Table 5.6, it can be concluded that COA performs better than other algorithms in terms of total minimum generation cost and computational time.

Table 5.6 Comparison of minimum total generation cost for test system-2

Optimization methods	Minimum total generation cost		Execution time in seconds
	$	`	
MPSO (Gillella et al. 2008)	15570.19	1003654.45	NA
GA (Gaing 2003)	15459.00	996487.14	NA
LAM-CON (Giulio Binetti et al. 2014)	15452.09	996041.72	NA
PSO (Gaing 2003)	15450.00	995907.00	NA
NPSO (Immanuel & Thanushkodi 2007)	15450.00	995907.00	NA
LAM-ITR (Wood & Wollenberg 1996)	15449.90	995900.55	NA
DHS (Ling Wang & Ling-po Li 2013)	15449.89	995899.91	NA
Binary Genetic Algorithm (BGA)	15449.34	995864.46	11.43
Modified PSO (MPSO)	15447.03	995715.55	17.64
MPSO (Cheng et al. 2008)	15447.00	995713.62	NA
Self adaptive DE (SADE)	15446.72	995695.57	8.10
COA	15443.32	995476.41	3.84

*NA-Not available

From the Table 5.6, it may be observed that, the proposed COA obtains minimum total generation cost as ` 995476.41. It can be observed that COA obtains a savings of ` 8178.04, ` 1010.73, ` 565.31, ` 430.59, ` 430.49, ` 424.14, ` 423.50, ` 388.05, ` 239.14, ` 237.21, and ` 219.16, more than when compared with MPSO, GA, LAM-CON, PSO, NPSO, LAM-ITR, DHS, BGA, MPSO, MPSO and SADE.

Table 5.7 Generation of electrical power output (MW) and minimum total generation cost for test system-2

Unit power Output(MW)		LAM-CON (Giulio Binetti et al. 2014)	DHS (Ling Wang & Ling-po Li 2013)	MPSO (Cheng et al. 2008)	MPSO	SADE	COA
P_1		449.31	447.53	446.71	448.47	444.12	443.89
P_2		173.16	173.28	173.01	171.00	171.77	173.78
P_3		266.13	263.48	265.00	263.49	262.56	265.66
P_4		127.41	139.03	139.00	124.77	126.38	139.28
P_5		174.40	165.49	165.23	171.01	171.77	165.94
P_6		86.02	87.16	86.78	97.02	99.14	86.97
Transmission Loss (MW)		13.27	12.96	12.73	12.77	12.74	12.46
Total Electrical Power Output (MW)		1276.27	1275.96	1275.70	1275.77	1275.74	1275.46
Minimum total generation cost	$	15452.09	15449.89	15447.00	15447.03	15446.72	15443.32
	`	995907.00	995899.91	995713.62	995715.55	995695.57	995476.41

Table 5.7 shows that the generation of electrical power output of each generator and minimum total generation cost of various recent methods. From the Table 5.7, summarizes that, the performance of COA is better than that of other algorithms in terms of minimum total generation cost and execution time.

Table 5.8 Statistical analysis of various methods for test system-2

Optimization Methods	Minimum total generation cost in $	Mean total generation cost in $	Maximum total generation cost in $	Standard deviation	Execution time in seconds
MPSO (Gillella et al. 2008)	15570.19	NA	NA	NA	NA
GA (Gaing 2003)	15459.00	15469.00	15524.00	0.06	NA
LAM-CON (Giulio Binetti et al. 2014)	15452.09	NA	NA	NA	NA
PSO (Gaing 2003)	15450.00	15454.00	15492.00	0.01	NA
NPSO (Immanuel & Thanushkodi 2007)	15450.00	15452.00	15454.00	NA	NA
LAM-ITR (Wood & Wollenberg 1996)	15449.90	NA	NA	NA	NA
DHS (Ling Wang & Ling-po Li 2013)	15449.89	15449.93	15449.99	0.76	NA
BGA	15449.34	15452.00	15473.00	2.70	11.43
MPSO	15447.03	15448.00	15469.00	2.16	17.64
MPSO (Cheng et al. 2008)	15447.00	15447.00	15455.00	2.53	7.58
SADE	15446.72	15446.80	15452.72	0.66	8.10
COA	15443.32	15443.00	15448.00	0.62	3.84

*NA- Not available

For the test system-2, statistical performance analyzed by conducting for 100 independent trail runs to the given power demand.

The results of minimum mean and maximum total generation cost, standard deviation and computational time were tabulated in Table 5.8.

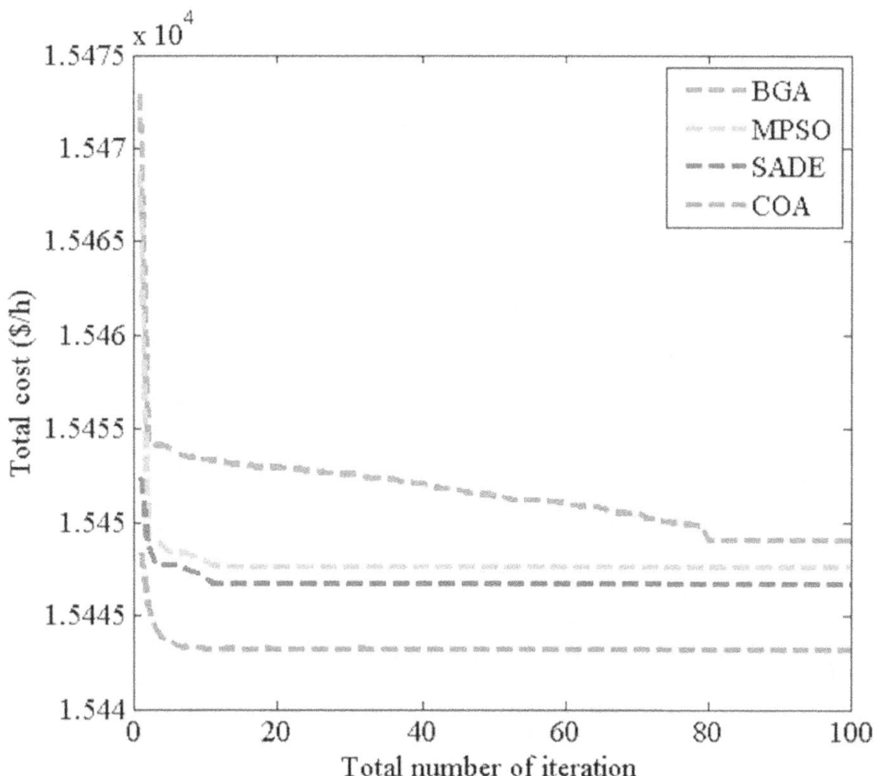

Figure 5.2 Convergence characteristics for test system-2

The convergence characteristic of COA for test system-2 is shown in Figure 5.2 along with well known methods of BGA, MPSO and SADE. Convergence characteristics have been plotted for 100 number of generations (Iteration) to total generation cost in $. The COA has been proven to have superior features, including high quality solutions in terms of total minimum generation cost, stable convergence characteristics and good computational efficiency

Table 5.9   Relative frequency of convergence in the range of total generation cost of various methods for test system-2

Range of total generation cost in $	15480.00-15460.00	15460.00-15450.00	15450.00-15448.00	15448.00-15447.50	15447.50-15444.00	15444.00-15443.50	15443.50-15443.20
BGA	1	73	26	-	-	-	-
MPSO	1	-	1	20	78	-	-
SADE	-	1	1	5	93	-	-
COA	-	-	1	1	2	2	94

The effectiveness of COA over the other well known optimization methods is established by attaining a total generation cost within the specific ranges out of 100 trail runs. To compare this, the results among BGA, MPSO, SADE and COA in a statistical manner, relative frequency of convergence in the range of total generating cost for test system-2 is tabulated in Table 6.9. From Table 5.8, it is observed that the successfulness of COA over other well known optimization methods is established in terms of high quality solution and better convergence.

5.4.3   Test System-3

This test sytem-3 consists of 15-generating units with 2, 5, 6 and 12 as prohibited operating zones in its input-output characteristics. The total demand for test system is 2630MW. The result obtained by COA is tabulated and compared in Table 5.10, 5.11 and 5.12 respectively.

Table 5.10   Comparison of minimum total generation cost of test system-3

Optimization Methods	Minimum total generation cost		Execution time in
	$	`	seconds
GA (Gaing 2003)	33113.00	2134463.98	NA
GA (Subbaraj et al. 2010)	33113.00	2134463.98	NA
PSO (Subbaraj et al. 2010)	32858.00	2118026.68	NA
PSO (Gaing 2003)	32858.88	2118083.40	NA
PAPSO (Subbaraj et al. 2010)	32715.67	2108852.09	NA
PSO-MSAF (Subbaraj et al 2010)	32713.09	2108685.78	NA
SARGA (Subbaraj et al. 2009)	32709.63	2108462.75	NA
ABC (Hemamalini et al. 2010)	32707.90	2108354.46	NA
HPSO-TVAC (Basu 2015)	32683.86	2106801.62	NA
DHS (Ling Wang & Ling-po Li 2013)	32588.92	2100681.78	NA
BGA	32651.17	2104694.42	26.40
MPSO	32571.56	2099562.76	13.20
LAM-CON (Giulio Binetti et al. 2014)	32568.54	2099368.09	NA
ESO (Pereira-Neto et al. 2005)	32568.54	2099368.09	NA
SADE	32556.41	2098586.19	11.60
COA	32551.99	2098301.28	3.97

*NA-Not available

From the Table 5.10, it may be observed that, the COA obtains minimum total generation cost as ` 2098301.28. It can be observed that COA obtains a savings of ` 36162.70, ` 36162.70, ` 19725.40, ` 19782.12, ` 10550.81, ` 10384.50, ` 10384.50, ` 10161.47, ` 10046.73, ` 8500.34, ` 2380.50, ` 6393.14, ` 1261.48, ` 1066.81, ` 1066.81 and ` 284.91, more than when compared with GA, PSO, PAPSO, PSO-MASF, SARGA, HPSO-TVAC, DHS, BGA, MPSO, LAM-CON, ESO, and SADE respectively. From the Table 5.10 it observed that COA performs better than other methods in terms of total minimum generation cost and computational time.

Table 5.11 Generation of electrical power output (MW) and minimum total generation cost for test system-3

Unit power Output (MW)		SARGA (Subbaraj et al. 2009)	LAM-CON (Binetti et al. 2014)	DHS (Ling Wang & Ling-po Li 2013)	SADE	COA
P_1		455.00	455.00	455.00	454.62	455.00
P_2		379.00	455.00	420.00	454.62	455.00
P_3		129.99	130.00	130.00	128.59	130.00
P_4		130.00	130.00	130.00	129.62	130.00
P_5		165.73	298.23	270.00	240.89	237.31
P_6		460.00	460.00	460.00	459.62	460.00
P_7		430.00	465.00	430.00	464.62	465.00
P_8		60.00	60.00	60.00	60.00	60.00
P_9		89.79	25.00	25.00	26.05	25.00
P_{10}		144.79	25.00	62.98	26.13	25.00
P_{11}		79.99	44.96	80.00	78.18	80.00
P_{12}		80.00	56.44	80.00	79.22	79.99
P_{13}		25.00	25.00	25.00	25.00	25.00
P_{14}		15.00	15.00	15.00	15.00	15.00
P_{15}		15.00	15.00	15.002	15.00	15.00
Transmission Loss (MW)		30.28	29.60	27.98	27.36	27.29
Total electrical power output (MW)		2660.28	2659.60	2657.98	2657.16	2657.29
Minimum total generation	$	32568.06	32568.54	32588.92	32556.41	32551.99

cost ($/h)	`	2108462.75	2099368.09	2100681.78	2098586.19	2098301.28

Table 5.11 gives the generation of electrical power output of each individual generator and minimum total generation cost of various recent methods. From the table 5.11, it is noted that, the performance of COA is better than that of other methods in terms of minimum total generation cost.

For the test system-3, statistical performance is analyzed by conducting 100 independent trail runs to the given power demand. The results of minimum mean and maximum total generation cost, standard deviation and computational time were tabulated in Table 5.12.

Table 5.12 Statistical analysis of various methods for test system-3

Optimization methods	Minimum total generation cost in $	Mean total generation cost in $	Maximum total generation cost in $	Standard Deviation	Execution time in seconds
GA (Gaing 2003)	33113.00	33228.00	33337.00	0.01	NA
GA (Subbaraj et al. 2010)	33113.00	33569.80	33756.23	9.21	NA
PSO (Subbaraj et al. 2010)	32858.00	32896.50	32987.05	6.42	NA
PSO (Gaing 2003)	32858.90	33039.00	33331.00	0.01	NA
PAPSO (Subbaraj et al. 2010)	32715.70	32789.30	32852.14	NA	NA
PSO-MSAF (Subbaraj et al. 2010)	32713.10	32759.60	32798.25	NA	NA
SARGA (Subbaraj et al. 2009)	32709.60	32730.80	32709.63	2.06	NA

Table 5.12 (Continued)

Optimization methods	Minimum total generation cost in $	Mean total generation cost in $	Maximum total generation cost in $	Standard Deviation	Execution time in seconds
ABC (Hemamalini et al. 2010)	32707.90	32708.00	32708.27	NA	NA
HPSO-TVAC (Basu 2015)	32683.90	32684.10	32684.71	NA	NA
DHS (Ling Wang & Ling-po Li 2013)	32588.90	32588.90	32588.94	2.30	NA
BGA	32651.20	32672.60	32960.33	13.62	26.40
MPSO	325720	32633.40	33101.12	10.34	13.20
ESO (Nasimul & Hitoshi Iba 2008)	32640.90	32710.00	32620.00	NA	NA
DE (Nasimul & Hitoshi Iba 2008)	32588.90	32641.40	32609.85	NA	NA
LAM-CON (Giulio Binetti et al. 2014)	32568.50	NA	NA	NA	NA
ESO (Pereira-Neto et al. 2005)	32568.50	32620.00	32710.00	NA	NA
SADE	32556.40	32575.90	32815.15	8.7	11.67
COA	32551.99	32562.70	32885.39	4.46	3.97

*NA- Not available

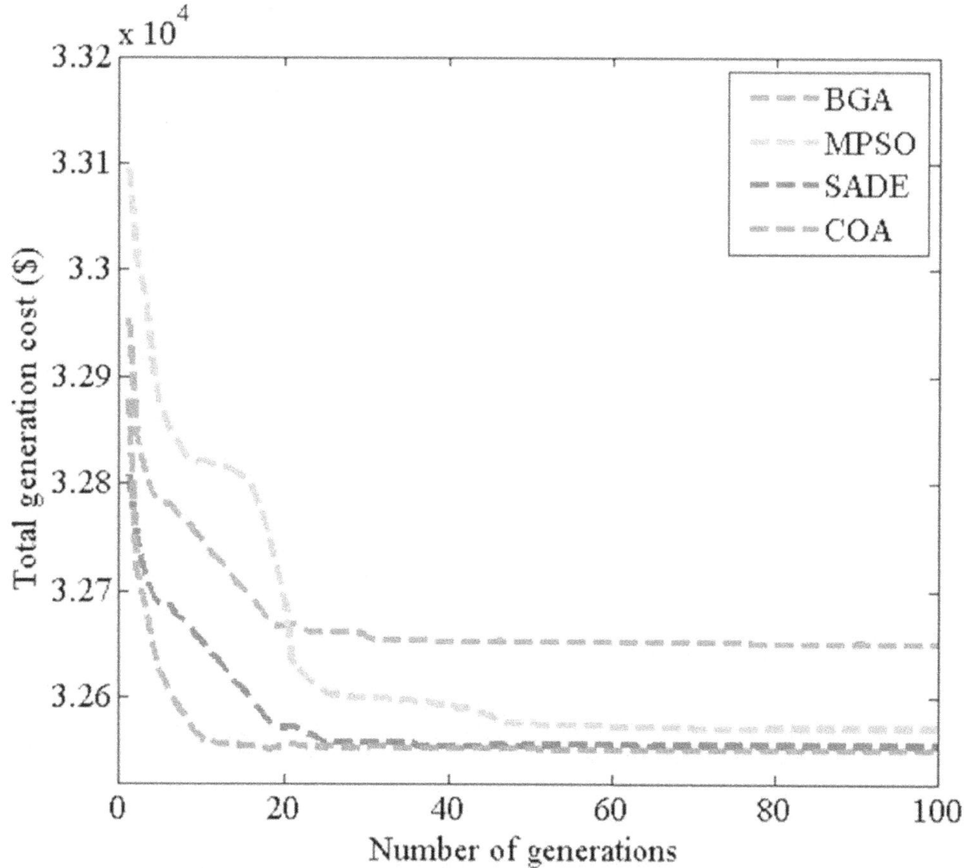

Figure 5.3 Convergence diagram for test system-3

The convergence characteristic COA for test system-3 is demonstrated in Figure 5.3 along with well known methods of BGA, MPSO and SADE. Convergence characteristics have been plotted for 100 number of generations (Iteration) to total generation cost in $. The proposed COA has been proven to have superior features, including high quality solutions in terms of total minimum generation cost, stable convergence characteristics and good computational efficiency.

Table 5.13   Relative frequency of convergence in the range of total generation cost of various methods for test system-3

Range of total generation cost in $	32885.00-32673.00	32673.00-32656.60	32656.60-32653.20	32653.20-32600.00	32600.00-32585.00	32585.00-32556.00	32556.00-32553.00	32553.00-32553.50	32553.50-32552.00
BGA	17	12	20	51	-	-	-	-	-
MPSO	20	3	2	2	18	55		-	-
SADE	7	2	-	8	10	18	55		-
COA	3	2	-	-	1	5	21	17	51

The effectiveness of COA over the other well known optimization methods is established by attaining a total generation cost within the specific ranges out of 100 trail runs. To compare this, the results among BGA, MPSO, SADE and COA in a statistical manner, relative frequency of convergence in the range of total generating cost for test system-3 is tabulated in Table 5.13. From Table 5.12, it is observed that the successfulness of COA over other well known optimization methods is established in terms of high quality solution and better convergence.

5.4.4   Test System-4

This test sytem-4 consists of 20-generating units. The thermal generator cost co-efficient, B-loss co-efficient and the total demand for test system is 2500MW. The result obtained by COA is tabulated and compared in Table 5.14, Table 5.15 and Table 5.16 respectively.

Table 5.14   Comparison of minimum total generation cost for test system-4

Optimization Method	Total Minimum generation cost		Execution time in seconds
	$	`	
BBO (Roy et al. 2014)	62457.55	4026013.67	NA
BGA	62456.80	4025965.33	29.28
MPSO	62456.80	4025965.33	48.21
SADE	62456.78	4025964.04	18.46
BBO (Anirudha et al. 2010)	62456.77	4025963.39	NA
LAM-ITR (Su et al. 2000)	62456.64	4025955.01	NA
HNN (Su et al. 2000)	62456.64	4025955.01	NA
ORCSA (Thang & Dieu 2015)	62456.64	4025955.01	NA
COA	62456.63	4025954.37	0.047

*NA- Not available

From the Table 5.14, it is observed that, COA obtains minimum total generation cost as ` 4025954.37. It can be observed that COA obtains a savings of ` 59.30, ` 10.96, ` 10.96, ` 9.67, ` 9.02, and ` 8.38 more than when compared with BGA, MPSO, SADE, BBO, LAM-ITR, HNN, and ORCSA. From Table 5.14, COA performs better than other algorithms in terms of total minimum generation cost and execution time.

Table 5.15   **Generation of electrical power output (MW) and minimum total generation cost for test sytem-4**

Unit power Output (MW)		BBO (Anirudha et al. 2010)	HNN (Su et al. 2000)	ORCSA (Thang & Dieu 2015)	COA
P_1		513.06	512.78	512.78	512.78
P_2		170.66	169.10	169.11	169.10
P_3		126.48	126.89	126.88	126.89
P_4		103.19	102.87	102.86	102.87
P_5		114.00	113.64	113.68	113.68
P_6		74.512	73.57	73.57	73.57
P_7		115.31	115.29	115.28	115.29
P_8		116.70	116.40	116.39	116.40
P_9		110.75	100.41	100.41	100.41
P_{10}		106.26	106.03	106.05	106.03
P_{11}		150.31	150.24	150.24	150.24
P_{12}		291.12	292.76	292.78	292.76
P_{13}		119.36	119.12	119.12	119.12
P_{14}		30.98	30.83	30.84	30.83
P_{15}		115.72	115.81	115.82	115.81
P_{16}		36.16	36.25	36.26	36.25
P_{17}		67.18	66.86	68.85	66.86
P_{18}		88.20	87.97	87.96	87.97
P_{19}		101.05	100.81	100.79	100.80
P_{20}		51.09	54.35	54.31	54.31
Transmission Loss (MW)		92.04	91.97	91.97	91.97
Total electrical power output (MW)		2592.04	2591.97	2591.97	2591.97
Minimum total generation cost ($/h)	$	62456.77	62456.64	62456.64	62456.63
	`	4025963.39	4025955.01	4025955.01	4025954.37

.   Table 5.15 lists the generation output of each individual generator and minimum total generation cost of various recent methods. From the Table 6.15, it is noted that, the performance of COA is better than that of other methods in terms of minimum total generation cost. For the test system-4, statistical performance is analyzed by conducting 100 independent trail runs to the given power demand. The results of minimum, mean and maximum total generation cost, standard deviation and computational time were tabulated in Table 5.16.

Table 5.16 Statistical analysis of various methods for test system-4

Optimization methods	Minimum total generation cost in $	Mean total generation cost in $	Maximum total generation cost in $	Standard deviation	Execution time in seconds
BBO (Roy et al 2014)	62457.55	NA	NA	NA	NA
BBO (Aniruddha & Chattopadhyay 2010)	62456.77	62456.80	62456.80	NA	NA
BGA	62456.80	62459.10	62457.00	0.052	42.16
MPSO	62456.80	62456.80	62457.00	0.035	48.21
SADE	62456.78	62456.80	62456.80	0.035	18.46
LAM-ITR (Su & Lin 2000)	62456.64	NA	NA	NA	NA
HNN (Leandro dos & Lee 2008)	62456.64	NA	NA	NA	NA
ORCSA (Thang & Dieu 2015)	62456.64	NA	NA	NA	NA
COA	62456.63	62456.70	62456.7	0.047	5.83

*NA- Not available

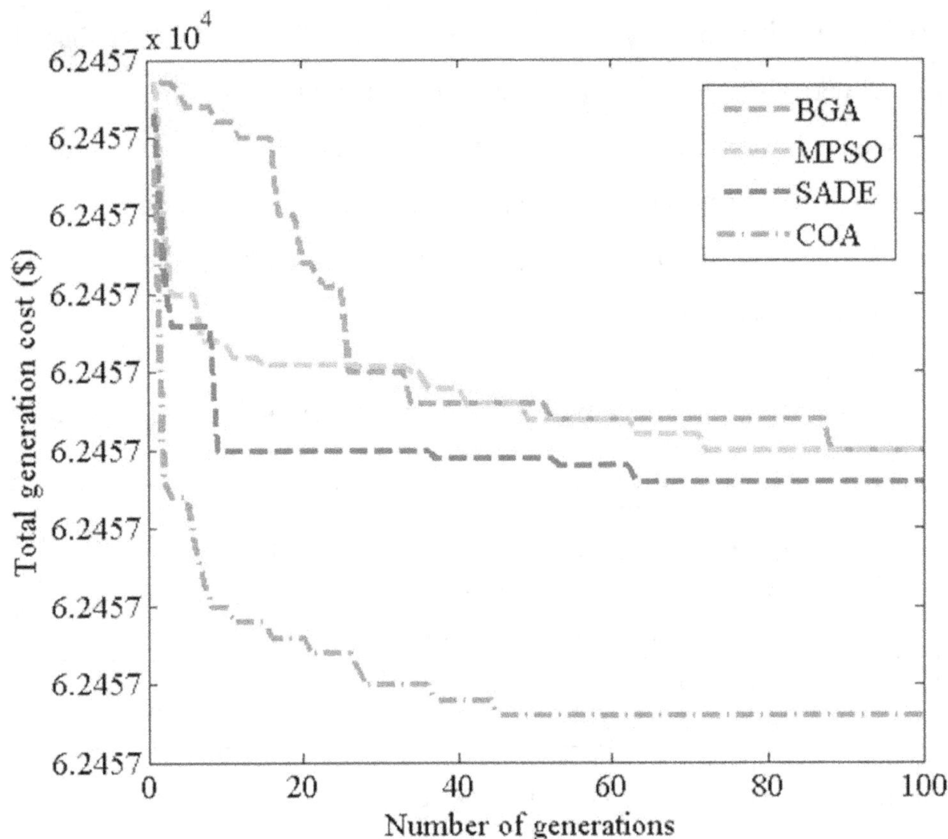

Figure 5.4 Convergence characteristics for test system-4

The convergence characteristic of COA for Test System-4 is shown in Figure 5.4 along with well known methods of BGA, MPSO and SADE. Convergence characteristics have been plotted for 100 number of generations to total generation cost in $. The COA has been proven to have superior features, including high quality solutions in terms of total minimum generation cost, stable convergence characteristics and good computational efficiency.

Table 5.17   Relative frequency of convergence in the range of total generation cost of various methods for test system-4

Range of total generation cost in $	62457.05-62456.88	62456.88-62456.80	62456.80-62456.79	62456.79-62456.78	62456.78-62456.68	62456.68-62456.64	62456.64-62456.62
BGA	25	62	13	-	-	-	-
MPSO	6	56	38	-	-	-	-
SADE	8	44	10	38	-	-	-
COA	-	-	-	1	14	21	64

The effectiveness of COA over the other well known optimization methods is established by attaining a total generation cost within the specific ranges out of 100 trail runs. To compare this, the results among BGA, MPSO, SADE and COA in a statistical manner, relative frequency of convergence in the range of total generating cost for test system-4 is tabulated in Table 5.17. From Table 5.16, it is observed that the successfulness of COA over other well known optimization methods is established in terms of high quality solution and better convergence.

5.5   CONVENTIONAL ED PROBLEM WITH WES

The COA is successfully employed for solving conventional ED problem with WES (EDWES) of test system-5. The simulation is carried out for 100 trail runs for the population size $M = 100$. The minimum fuel cost, maximum cost, mean cost, standard deviation and the execution times are used to compare the performance of COA with other methods.

5.5.1   Test System-5

The test system-5 is a 3-generator system with the total demand is set to 300MW. The results obtained for the test system-5 are   tabulated in Table 5.18. In this simulation of the rated output of WES is 100MW.

Table 5.18 lists the generation output of each individual generator and minimum total generation cost. The COA obtained optimal result of ` 182756.35. It can be observed that COA obtains a savings of ` 105.71, ` 6.04, and ` 50.28, more than when compared with BGA, MPSO, and SADE.

From the Table 5.18, it is noted that, the performance of COA is better than that of other methods in terms of minimum total generation cost. For the test system-5 with WES, the performance analyzed for   100 independent trail runs to the given power demand.

Table 5.18   Generation of electrical power output (MW) and minimum total generation cost for test system-5

Unit power output (MW)		BGA	MPSO	SADE	COA
P_1		151.83	152.01	152.90	153.65
P_2		38.01	38.01	38.01	38.10
P_3		16.94	16.74	15.75	15.00
P_{wes}		98.02	98.02	98.02	98.02
Transmission Loss (MW)		4.79	4.77	4.66	4.58
Total Electrical Power Output (MW)		304.79	304.77	304.66	304.58
Minimum total generation cost ($/h)	$	2836.83	2836.68	2835.97	2835.19
	`	182862.06	182852.39	182806.63	182756.35

5.5.2 Test System-6

This test sytem-6 consists of 6-generating units with prohibited operating zones with WES. The total demand for test system is 1263MW. The results obtained for minimum total generation cost is tabulated and compared in Table 5.19, Table 5.20 and Table 5.21 respectively. In this simulation of the rated output of WES is 500MW.

Table 5.19 Comparison of minimum total cost of test system-6

Optimization method	Minimum total generation cost		Execution time in seconds
	$	`	
BGA	10494.95	676504.48	11.23
MPSO	10493.72	676425.19	17.84
SADE	10493.66	676421.32	8.43
COA	10488.20	676069.37	3.84

From the Table 5.19, it is observed that, the COA obtains minimum total generation cost as ` 676069.37. It can be observed that COA obtains a savings of ` 435.11, ` 355.82, and ` 351.95, is more than when compared with BGA, MPSO and SADE. From the Table 5.19, it may be noted that, the performance of COA is better than that of other methods in terms of minimum total generation cost and execution time.

Table 5.20 shows the generation output of each individual generator and minimum total generation cost and the COA has a better performance when compared with other methods.

For the test system-6 with WES, statistical performance is analyzed by conducting 100 independent trail runs to the given power demand. The results of minimum, mean and maximum total generation cost, standard deviation and computational time were tabulated in Table 5.21.

Table 5.20   Generation of electrical power output (MW) and minimum total generation cost for test sytem-6

Unit power Output (MW)		MPSO	BGA	SADE	COA
P_1		327.12	334.82	332.99	334.46
P_2		77.44	72.17	106.87	89.83
P_3		171.25	176.70	177.99	176.53
P_4		57.44	52.17	50.00	50.10
P_5		97.44	92.17	60.27	77.13
P_6		47.44	50.00	50.00	50.00
P_{WES}		490.09	490.09	490.09	490.09
Transmission Loss (MW)		5.23	5.12	5.21	5.13
Total Electrical Power Output (MW)		1268.23	1298.12	1268.21	1268.13
Minimum total generation cost	$	10494.95	10493.72	10493.66	10488.20
	`	676504.48	676425.19	676421.32	676069.37

Table 5.21 Statistical analysis of various methods for test system-6

Optimization methods	Minimum total generation cost in $	Mean total generation cost in $	Maximum total generation cost in $	Standard Deviation	Execution time in seconds
BGA	10495.00	10496.10	10517.80	3.929	11.23
MPSO	10493.70	10496.90	10513.60	2.708	17.84
SADE	10493.70	10493.90	10501.60	2.172	8.43
COA	10488.20	10488.90	10499.72	0.856	3.84

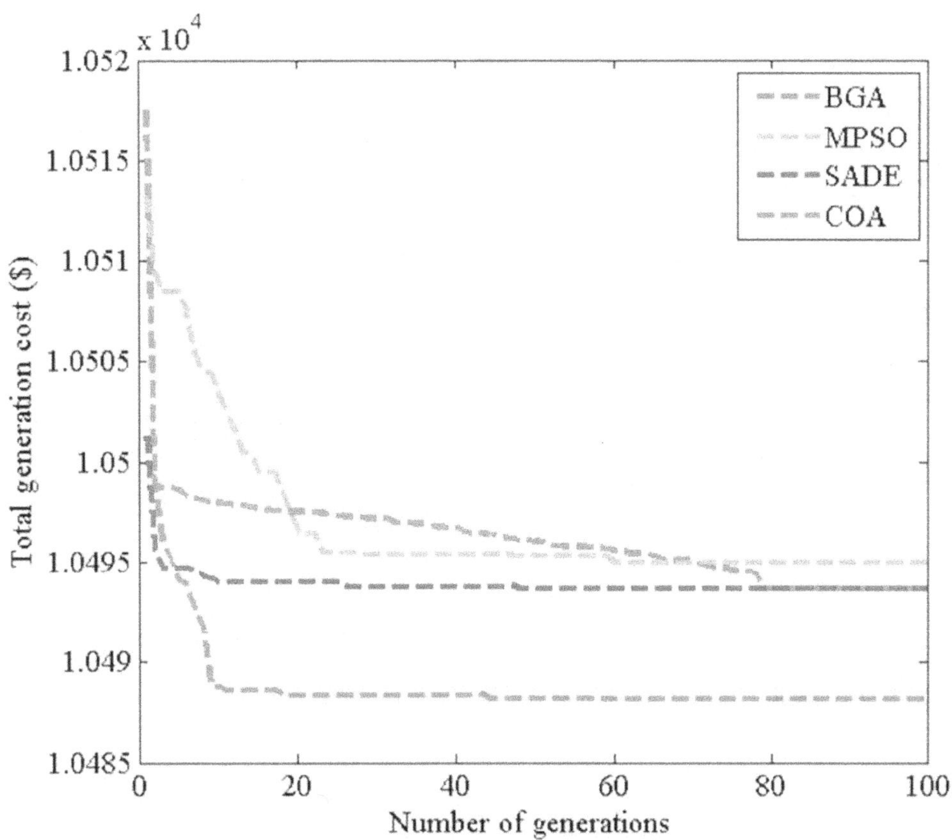

Figure 5.5 Convergence characteristics for test system-6

The convergence characteristic of proposed COA for Test System-6 with WES is shown in Figure 5.5 along with well known methods of BGA, MPSO and SADE. Convergence diagram have been plotted for 100 number of generation (iteration) with total generating cost in $/h. Convergence of COA has faster and superior convergence rate compared to BGA, MPSO and SADE.

Table 5.22   Relative frequency of convergence in the range of total generation cost of various methods for test system-6

Range of total generation cost in $	10517.80-10496.50	10496.50-10495.50	10495.50-10495.30	10495.30-10494.20	10494.20-10493.80	10493.80-10493.75	10493.75-10491.00	10491.00-10488.40	10488.40-10488.20
BGA	31	9	4	34	22	-	-	-	-
MPSO	21	5	32	42	-	-	-	-	-
SADE	2	-	-	-	23	22	53	-	-
COA	2	-	-	-	1	1	3	35	58

The effectiveness of COA over the other well known optimization methods is established by attaining a total generation cost within the specific ranges out of 100 trail runs.

To compare this, the results among BGA, MPSO, SADE and COA in a statistical manner, relative frequency of convergence in the range of total generating cost for test system-6 with WES is tabulated in Table 5.22. From Table 5.21, it is observed that the successfulness of COA over other well known optimization methods is established in terms of high quality solution and better convergence.

5.5.3   Test System-7

This test sytem-7 consists of 15-generating units with WES. The total demand for test system is 2630MW. In this simulation of the rated output of WES is 455MW. The result obtained by COA is tabulated in Table 5.23, Table 5.24 and Table 5.25 respectively.

Table 5.23 Comparison of minimum total cost of test system-7

Optimization Methods	Minimum cost ($/h)		Execution time in seconds
	$	`	
BGA	29510.75	1902262.95	28.40
MPSO	29509.82	1902203.00	14.20
SADE	29509.40	1902175.92	11.67
COA	29507.50	1902053.45	3.40

From the Table 5.23, it is observed that, COA obtains minimum total generation cost as ` 1902053.45. It can be observed that COA obtains a savings of ` 209.49, ` 149.55, and ` 122.47 are more than when compared with BGA, MPSO and SADE.

From the Table 5.23, it may be noted that, the performance of COA is better than that of other methods in terms of minimum total generation cost and computational time. Table 5.24 lists the generation output of each individual generator and minimum total generation cost. COA performance better compared with other methods.

Table 5.24 Generation of electrical power output (MW) and minimum total generation cost for test sytem-7

Unit power Output (MW)		BGA	MPSO	SADE	COA
P_1		356.01	357.01	356.10	356.10
P_2		267.15	267.15	269.15	267.15
P_3		129.99	129.99	129.99	129.99
P_4		129.99	129.99	129.99	129.99
P_5		149.99	149.99	149.99	149.99
P_6		459.99	459.99	459.99	459.99
P_7		464.99	464.99	464.99	464.99
P_8		62.70	61.70	59.99	59.99
P_9		25.00	25.00	25.00	24.99
P_{10}		25.00	25.00	25.00	24.99
P_{11}		31.46	31.46	31.19	31.20
P_{12}		45.20	45.20	46.20	47.99
P_{13}		25.00	25.00	25.00	25.00
P_{14}		15.00	15.00	15.00	15.00
P_{15}		15.00	15.00	15.00	15.00
P_{wes}		445.98	445.98	445.98	445.98
Transmission Loss (MW)		18.47	18.45	18.39	18.36
Total electrical power output (MW)		2648.47	2648.45	2648.39	2648.31
Minimum total generation cost ($/h)	$	29510.75	29509.82	29509.40	29507.50
	`	1902262.95	1902203.00	1902175.92	1902053.45

For the test system-7 with WES, the statistical performance analyzed by conducting 100 independent trail runs to the given power demand. The results of minimum, mean and maximum total generation cost, standard deviation and computational time were tabulated in Table 5.25. The total minimum cost of proposed COA is better than that of other methods is tabulated in the Table 5.25.

Table 5.25 Statistical analysis of various methods for test system-7

Optimization Methods	Minimum total generation cost in $	Mean total generation cost in $	Maximum total generation cost in $	Standard deviation	Execution time in seconds
BGA	29510.80	29630.10	29898.80	109.93	28.40
MPSO	29509.80	29568.40	29789.00	92.31	14.20
SADE	29509.40	29531.97	29872.80	58.93	11.67
COA	29507.50	29523.99	29708.56	38.59	5.40

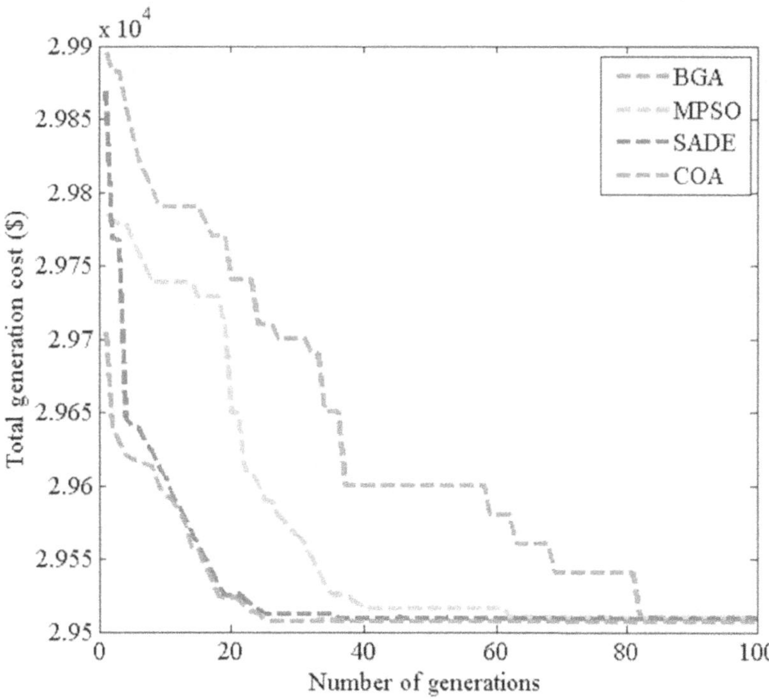

Figure 5.6 Convergence characteristics for test system-7

The convergence characteristic of COA for Test System-7 with WES is shown in Figure 5.6 along with well known methods of BGA, MPSO and SADE. Convergence diagram have been plotted for 100 number of generation with total generating cost in $/h. Convergence of COA faster and superior convergence rate compared to GA, PSO and DE.

Table 5.26   Relative frequency of convergence in the range of total generation cost of various methods for test system-7

Range of total generation cost in $	29899.80-29794.00	29794.00-29601.00	29601.26-29518.40	29518.40-29512.42	29512.42-29509.00	29509.00-29507.32
BGA	8	20	45	8	19	-
MPSO	-	23	16	22	39	-
SADE	1	8	13	13	65	-
COA	-	1	19	3	9	68

The effectiveness of COA over the other well known optimization methods is established by attaining a total generation cost within the specific ranges out of 100 trail runs. To compare this, the results among BGA, MPSO, SADE and COA in a statistical manner, relative frequency of convergence in the range of total generating cost for test system-7 with WES is tabulated in Table 5.26. From Table 5.25, it is observed that the successfulness of COA over other well known optimization methods is established in terms of high quality solution and better convergence.

5.5.4   Test System-8

This test sytem-8 consists of 20-generating units with WES. The total demand for test system is 2500MW. In this simulation of the rated output of WES is 600MW. The result obtained by COA is tabulated in Table 5.27, Table 5.28 and Table 5.29 respectively.

Table 5.27 Comparison of minimum total cost for test system-8

Optimization Methods	Minimum cost ($/h)		Execution time in seconds
	$	`	
BGA	54330.85	3502166.59	26.40
MPSO	54330.72	3502158.21	13.20
SADE	54330.29	3502130.49	11.67
COA	54328.27	3502000.28	3.97

From the Table 5.27, it is observed that, COA obtains minimum total generation cost as ` 3502000.28. It can be observed that COA obtains a savings of ` 166.31, ` 157.93, and ` 130.21 more than when compared with BGA, MPSO and SADE.

From the Table 5.27, it may be noted that, the performance of COA is better than that of other methods in terms of minimum total generation cost and computational time.

Table 5.28   Generation of electrical power output (MW) and minimum total generation cost for test sytem-8

Unit power Output (MW)		BGA	MPSO	SADE	COA
P_1		370.41	370.4	377.5	375.86
P_2		122.92	124.92	124.96	124.96
P_3		74.20	74.2	74.2	73.20
P_4		73.99	73.99	73.99	73.99
P_5		81.42	81.42	81.42	81.42
P_6		51.31	51.31	51.31	51.31
P_7		103.74	103.74	103.74	102.74
P_8		67.89	67.89	67.89	67.84
P_9		74.58	74.58	74.58	74.50
P_{10}		62.18	62.18	62.18	61.03
P_{11}		143.48	143.48	143.48	145.88
P_{12}		263.12	263.12	263.12	264.12
P_{13}		97.05	97.05	97.05	105.15
P_{14}		45.03	45.03	39.03	36.23
P_{15}		96.40	96.4	96.4	96.93
P_{16}		37.15	35.15	34.15	33.15
P_{17}		39.78	39.78	39.78	39.78
P_{18}		57.49	57.49	57.49	57.49
P_{19}		72.65	72.65	72.65	72.65
P_{20}		37.72	37.72	37.72	34.72
P_{wes}		588.11	588.11	588.11	588.11
Transmission Loss (MW)		60.62	60.62	60.75	61.07
Total electrical power output (MW)		2560.62	2560.62	2560.75	2561.07
Minimum total generation cost ($/h)	$	54330.85	54330.72	54330.29	54328.27
	`	3502166.59	3502158.21	3502130.49	3502000.28

Table 5.28 shows the generation output of each individual generator and minimum total generation cost and COA has a better performance when compared with other methods.

For the test system-8 with WES, statistical performance is analyzed by conducting 100 independent trail runs to the given power demand. The results of minimum mean and maximum total generation cost, standard deviation and computational time were tabulated in Table 5.29.

Table 5.29 Statistical analysis of various methods for test system-8

Optimization Methods	Minimum total generation cost in $	Mean total generation cost in $	Maximum total generation cost in $	Standard deviation	Execution time in seconds
BGA	54330.85	54351.02	54412.60	15.43	26.40
MPSO	54330.72	54348.44	54401.96	16.01	13.20
SADE	54330.29	54334.67	54398.64	12.10	11.67
COA	54328.27	54330.34	54385.62	7.42	3.97

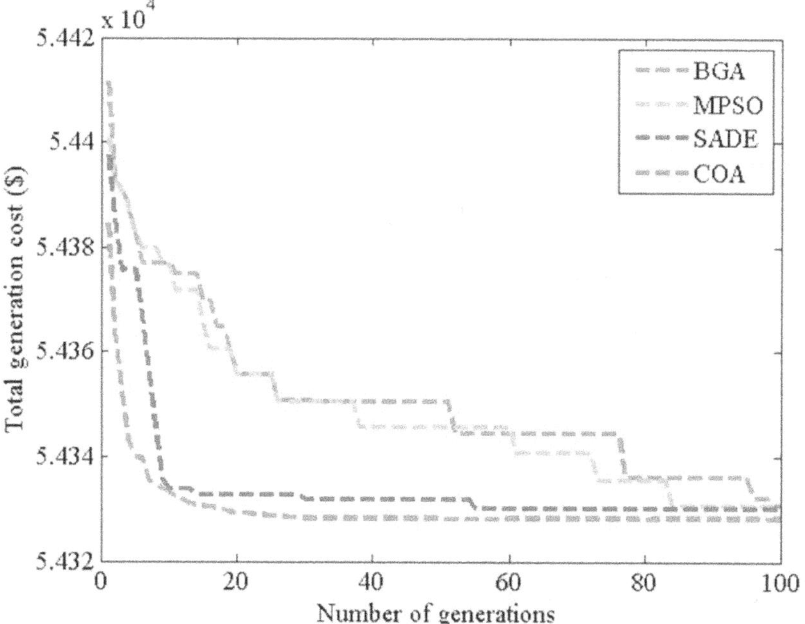

Figure 5.7 Convergence characteristics for test system-8

The convergence characteristic of COA for Test System-8 with WES is shown in Figure 5.7 along with well known methods of BGA, MPSO and SADE. Convergence diagram have been plotted for 100 number of generation (iteration) with total generating cost in $/h. Convergence of COA faster and superior convergence rate compared to BGA, MPSO and SADE.

Table 5.30   Relative frequency of convergence in the range of total generation cost of various methods for test system-8

Range of total generation cost in $	54412.60-54392.99	54392.99-54385.00	54385.00-54352.53	54352.53-54332.85	54332.85-54331.85	54331.85-54330.22	54330.22-54328.27
BGA	2	2	21	73	2	-	-
MPSO	2	3	20	58	2	15	-
SADE	-	1	4	16	33	46	-
COA	-	-	2	8	2	4	84

The effectiveness of COA over the other well known optimization methods is established by attaining a total generation cost within the specific ranges out of 100 trail runs. To compare this, the results among BGA, MPSO, SADE and COA in a statistical manner, relative frequency of convergence in the range of total generating cost for test system-8 with WES is tabulated in Table 5.30. From Table 5.29, it is observed that the successfulness of COA over other well known optimization methods is established in terms of high quality solution and better convergence.

Table 5.31 shows the comparison of optimal total generation cost achieved by COA technique for conventional ED problem with and without WES.

Table 5.31 Optimal generation output obtained by COA for ED and EDWES problems for various test systems considered

Units in MW	COA							
	3 Generating units		6 Generating units		15 Generating units		20 Generating units	
	Without wind	With wind	Without wind	With wind	Without wind	With wind	Without wind	With wind
P_1	214.61	153.65	443.890	334.46	455.00	356.10	512.78	375.86
P_2	78.63	38.10	173.78	89.83	455.00	267.15	169.11	124.96
P_3	15.93	15.00	265.66	176.53	130.00	129.99	126.89	73.20
P_4	-	-	139.28	50.10	130.00	129.99	102.87	73.99
P_5	-	-	165.94	77.13	237.31	149.99	113.68	81.42
P_6	-	-	86.97	50.00	460.00	459.99	73.57	51.31
P_7	-	-	-	-	465.00	464.99	115.29	102.74
P_8	-	-	-	-	60.00	59.99	116.39	67.84
P_9	-	-	-	-	25.00	24.99	100.41	74.50
P_{10}	-	-	-	-	25.00	24.99	106.03	61.03
P_{11}	-	-	-	-	80.00	31.20	150.24	145.88
P_{12}	-	-	-	-	79.99	47.99	292.76	264.12
P_{13}	-	-	-	-	25.00	24.99	119.12	105.15

Table 5.31   (Continued)

Units in MW	COA							
	3 Generating units		**6 Generating units**		**15 Generating units**		**20 Generating units**	
	Without wind	**With wind**	**Without wind**	**With wind**	**Without wind**	**With wind**	**Without wind**	**With wind**
P_{14}	-	-	-	-	15.00	14.99	30.83	36.23
P_{15}	-	-	-	-	15.00	14.99	115.81	96.93
P_{16}	-	-	-	-	-	-	36.25	33.15
P_{17}	-	-	-	-	-	-	66.86	39.78
P_{18}	-	-	-	-	-	-	87.97	57.49
P_{19}	-	-	-	-	-	-	100.80	72.65
P_{20}	-	-	-	-	-	-	54.31	34.72
TL (MW)	9.17	4.58	12.46	5.13	27.29	18.36	91.97	61.07
TP (MW)	309.17	304.58	1275.46	1268.13	2657.29	2648.31	2591.97	2561.07
Pw (MW)	-	98.018	-	490.09	-	445.9816	-	588.11
Total generation cost ($)	3609.225	2835.19	15443.32	10488.20	32551.99	29507.50	62456.63	54328.27
% Reduction	21.44%		32.08%		9.07%		13.01%	

From the Table 5.31, COA technique is implemented to the ED problem incorporating WES. The total number of thermal generating units considered is: 3, 6, 15 and 20. In addition, to the total number of thermal generating units in each test systems, one WES unit is considered whose power is generated based on the impact of wind turbulence with mean wind speed. WES is incorporated only for short duration and the wind speed variation effect is considered. The turbulence effect considered for calculation of wind power is 40% which is taken from literature.

5.6   CAPTURE OF MAXIMUM WIND ENERGY (CMWE)

The performance of COA is evaluated by solving CMWE of test system-9 in section 5.2. The parameter value of COA, grazing area $\therefore \varphi \vartheta$ at 0.35 is used for entire simulations, length of the rope $lr = 0.9376$ and the grazing angle $\alpha = 2.1658$. To verify the performance of COA the program is conducted for 100 trial runs for the population size M=100.

5.6.1   Test System-9

In this simulation COA is successfully applied without considering pitch angle of the wind turbine of test system-9. In addition the scale factor is 6m/s, shape factor 2m/s, cut-in speed as 2.5 m/s, cut-off speed 15 m/s, radius of the wind turbine is 30m, Multiplication factor of gear box is at 60, the air density taken as $\pi = 1.225 kgm^{-3}$, number of poles is 4, rated power as 2MW, nominal voltage 960 V, HVDC voltage 150kv, the rating of power transformers are 2.5MVA and the wind farm transformers rating are 10MVA. The simulation is carried out for capturing maximum wind energy for various wind speed 4m/s, 6m/s, 10m/s, 12m/s and 14m/s are tabulated in Table 5.32.

Table 5.32 lists a comparative study of the maximum capturing of power by COA with PSO. From Table 5.32 it is observed that the power capture is significantly increased by COA of 3.35%, 11.05%, 20.90%, 25.70%, 9.50% and 10.50% when compared with PSO.

Table 5.32 Comparison of maximum wind energy captured for 2MW rated WES

Wind speed m/sec	PSO (Oriol Gomis-Bellmunt et al. 2010)		COA	
	P_w in MW	% of wind power capture	P_w in MW	% of wind power capture
4	0.000	0.00	0.067	03.35
6	0.129	6.45	0.341	17.05
8	0.420	21.00	0.838	41.90
10	0.832	41.60	1.346	67.30
12	1.270	63.50	1.460	73.00
14	1.670	83.50	1.820	94.00

Table 5.32 gives the maximum power captured by COA, which analyzes the different wind speed scenarios of off shore wind farms. It clearly shows the percentage of capturing wind energy is improved from 83.5% to 94.0% for the 2MW rated WES by COA technique.

Figure 5.8 shows the simulated output of COA for different wind speed scenarios of a single wind turbine for a constant frequency of 50Hz. The simulation is carried out for capturing maximum wind energy for various wind speed 4m/s, 6m/s, 10m/s, 12m/s and 14m/s and the diagram has been plotted between tip speed ratios to generated wind power.

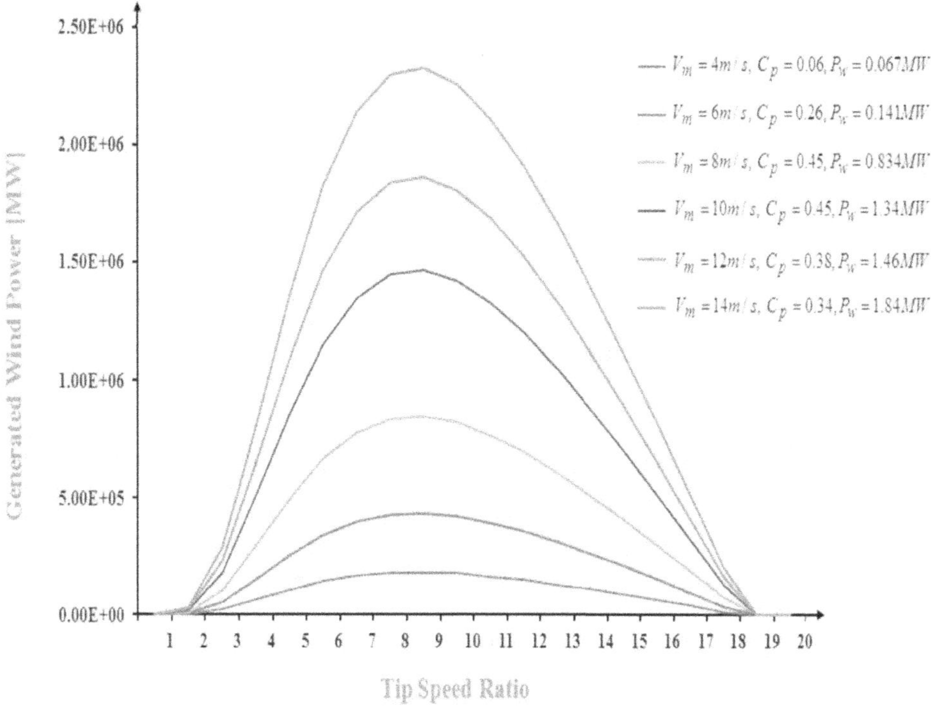

Figure 5.8 Power generated by SWT for different wind speed

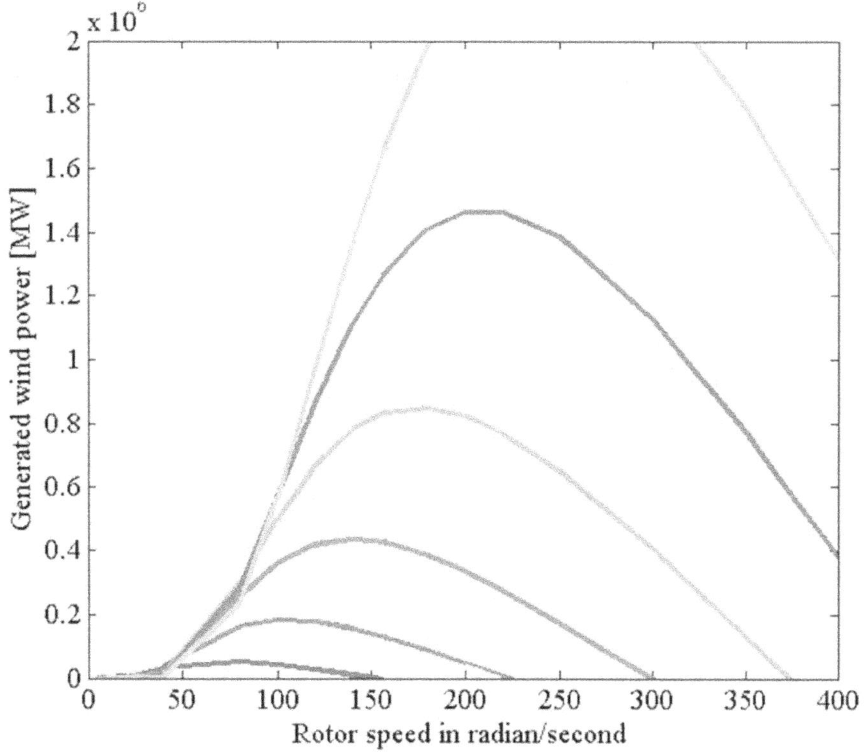

**Figure 5.9   Power generated by SWT for different wind speed vs rotor
speed**

Figure 5.9 shows the simulated output of COA for different wind speed scenarios of a single wind turbine for a constant frequency of 50Hz. The simulation is carried out for capturing maximum wind energy for various wind speed 4m/s, 6m/s, 10m/s, 12m/s and 14m/s and the diagram has been plotted between rotor speeds to generated wind power.

5.7   CONCLUSION

This chapter presented the successful implementation of COA in ED, EDWES and CMWE to the test systems in section 5.2. From the simulation results it is observed that COA has the power of exploration and exploitation in attaining better convergence towards optimal solutions for the considered test system. The performance of COA is also statistically analyzed and compared with various existing / recently reported algorithms in terms of maximum, minimum and mean total generation cost, standard deviation and execution time of ED and EDWES problem for the considered test system. In addition, the performance of COA is also statistically analyzed and compared with existing algorithm in terms of percentage of capture and generated power of CMWE problem.

CHAPTER 6

CONCLUSION AND SCOPE FOR FUTURE WORK

6.1   CONCLUSION

This thesis has dealt with successful implementation of the newly developed Capra Optimization Algorithm (COA) based on bio-inspired behaviour of Capra. COA is implemented for solving ED, EDWES and CMWE problems for standard test systems. The ED, EDWES and CMWE problems involve highly non-linear real numbers leading to multiple local optima in the objective functions. Hence an effective and constraint handling optimization method is required to solve these non-linear problems. The proposed COA handles the discontinuities, non-linearities and feasible operating region involved in ED, EDWES and CMWE problems effectively.

The performance of the proposed COA is compared with other optimization methods for 9 test systems. These test systems are classified into following problems,

i)   ED problem incorporating without WES.

ii) ED problem incorporating with WES

iii) Capture of maximum power from WES.

COA is applied to solve the conventional ED problem with 3, 6, 15 and 20 generating units. The proposed COA has provided better solution compared to that of GA, T-NN, APSO, SARGA, LAM-ITR, BBO, DE/BBO,

MPSO, LAM-CON, NPSO, DHS, DE, PAPSO, HPSO-TVAC, PSO-MSAF, ESO and ABC reported in the literature. The performance evaluation criteria used for comparison of COA with other algorithms are solution quality, convergence characteristics and computational efficiency.

In addition, COA is applied to solve EDWES problem with 3, 6, 15 and 20 thermal generating units along with the addition of single WES on each test system. The proposed COA has provided better solution compared to that of GA, PSO and DE for considered test systems. The performance evaluation criteria used for comparison of COA with other methods are solution quality, convergence characteristics and computational efficiency.

Also, COA is applied to solve CMWE problem to standard test system consisting of SWT having constant frequency off shore wind farm. The proposed COA has provided enhanced solution in order to achieve maximum capture of power from SWT when compared to PSO for the considered test system.

To conclude, the performance of proposed COA is better than that of other various well-known methods to solve ED, EDWES and CMWE problems for the standard test systems considered.

6.2   FUTURE WORK

Further research work may be followed on the following area:

- It may be improved by further research work by self-adaptation may be introduced in COA and its performance may be analyzed on the parameters like grazing land type, length of the rope and bite rate.

- It can be used to solve other power system engineering problems like unit commitment, maintenance scheduling, power system planning, power system stability and voltage stability.

- COA may be utilized and implemented through hardware to solve the real world power system problems.

- There are issues that are yet to be addressed in CMWE. The tuning of controllers can be done using COA parameters for optimized control and thus enhancing the WES performance. The proposed COA may further be implemented for the online tuning of CMWE.

APPENDIX 1

GRAZING HERBIVORE SPECIES SELECTION FOR AN APPLICATION

Table A1.1 Grazing methods and selection of large grazing herbivores

Species	Biting method	Selective Ability	Minimum Sward Height	Particular preference
Sheep	Biting/Shearing. Lower incisor	Highly selective	3cm	Avoid mat-grass, rushes. Eat mosses only can't avoid. Wethers and Hebridean eat rough vegetation
Red deer	Biting/Shearing. Lower incisor	Selective	4cm	More liable to eat heather and trees than are sheep
Cattle	Wrap tongues and pull or Biting/Shearing. Lower incisor	Slightly Selective	>6cm	More liable to eat rough vegetation like mat grass and purple moor grass than sheep/deer
Goats	Biting/Shearing. Lower incisor	Highly selective	>6cm	Will eat a very varied diet. Will eat mat grass and in spring rushes.
Horses Ponies	Biting/Shearing. Lower incisor	Selective	2cm	Prefer vegetation with a high digestibility, even if the sward is very short

Table A1.1 (Continued)

Species	Biting method	Selective Ability	Minimum Sward Height	Particular preference
Hares	Biting/Shearing. Lower incisor	Very highly Selective	3cm	Prefer heather on mineral rich soils to that growing on poorer soils. Select grass from heather
Rabbits	Biting/Shearing. Lower incisor	Very highly Selective	1cm	Avoid aromatic, prickly, hairy, fibrous, toxic or low digestibility plant species

APPENDIX 2

TEST SYSTEM FOR CONVENTIONAL GENERATING UNITS (3-GENERATING UNITS WITHOUT VALVE POINT LOADING)

Table A2.1 Test system data for 3-generating unit with prohibited operating zone and ramp-rate limits

Units	$P_{tp,j}^{min}$	$P_{tp,j}^{max}$	$a_{tp,j}$	$b_{tp,j}$	$c_{tp,j}$	$P_{tp,j}^{0}$	UR_j	DR_j	Prohibited Zones (MW)
1	50	250	0.00525	8.663	328.13	215	55.0	95.0	[105,117] [165,177]
2	5	150	0.00609	10.04	136.91	72.0	55.0	78.0	[50,60] [92,102]
3	15	100	0.00592	9.76	59.16	98.0	45.0	64.0	[25,32] [60,67]

B-co-efficients of 3-generating units

$$B_{tp,ij} = \begin{bmatrix} 0.000136 & 0.0000175 & 0.000184 \\ 0.0000175 & 0.000154 & 0.000283 \\ 0.000184 & 0.000283 & 0.00161 \end{bmatrix}$$

APPENDIX 3

TEST SYSTEM FOR CONVENTIONAL GENERATING UNITS (6-GENERATING UNITS WITHOUT VALVE POINT LOADING)

Table A3.1 Test system data for 6-generating unit with prohibited operating zone and ramp-rate limits

Units	$P_{tp,j}^{\min}$	$P_{tp,j}^{\max}$	$a_{tp,j}$	$b_{tp,j}$	$c_{tp,j}$	$P_{tp,j}^{0}$	UR_j	DR_j	Prohibited Zones (MW)
1	100	500	0.0070	7.0	240	440	80	120	[210 240] [350 380]
2	50	200	0.0095	10.0	200	170	50	90	[90 110] [140 160]
3	80	300	0.0090	8.5	220	200	65	100	[150 170] [210 240]
4	50	150	0.0090	11.0	200	150	50	90	[80 90] [110 120]
5	50	200	0.0080	10.5	220	190	50	90	[90 110] [140 150]
6	50	120	0.0075	12.0	190	110	50	90	[75 85] [100 105]

B-co-efficients of 3-generating units

$$
B_{ij} = \begin{bmatrix}
0.0017 & 0.0012 & 0.0007 & -0.0001 & -0.0005 & -0.0002 \\
0.0012 & 0.0014 & 0.0009 & 0.0001 & -0.0006 & -0.0001 \\
0.0007 & 0.0009 & 0.0031 & 0.0000 & 0.0010 & -0.0006 \\
-0.0001 & 0.0001 & 0.0000 & 0.0024 & -0.0006 & -0.0008 \\
-0.0005 & -0.0006 & -0.0010 & -0.0006 & 0.0129 & -0.0002 \\
-0.0002 & -0.00001 & -0.0006 & -0.0008 & -0.00002 & 0.0150
\end{bmatrix}
$$

$$B_{oi} = 1.0e^{-03*}\begin{bmatrix} -0.3908 & -0.1297 & 0.7074 & 0.0591 & 0.2161 & -0.6635 \end{bmatrix}$$

$$B_{oo} = 0.056$$

APPENDIX 4

TEST SYSTEM FOR CONVENTIONAL GENERATING UNITS (15-GENERATING UNITS WITHOUT VALVE POINT LOADING)

Table A4.1 Test system data for 3-generating unit with prohibited operating zone and ramp-rate limits

Units	$P_{tp,j}^{min}$	$P_{tp,j}^{max}$	$a_{tp,j}$	$b_{tp,j}$	$c_{tp,j}$	$P_{tp,j}^{0}$	UR_j	DR_j	Prohibited Zones (MW)
1	150	455	0.000299	10.1	671	400	120	80	-
2	150	455	0.000183	10.2	574	300	120	80	[185 225] [305 335] [420 450]
3	20	130	0.001126	8.8	374	105	130	130	-
4	20	130	0.001126	8.8	374	100	130	130	-
5	150	470	0.000205	10.4	461	90	120	80	[180 200] [305 335] [390 420]
6	135	460	0.000301	10.1	630	400	120	80	[230 255] [365 395] [430 455]
7	135	465	0.000364	9.8	548	350	120	80	-
8	60	300	0.000338	11.2	227	95	100	65	-
9	25	162	0.000807	11.2	173	105	100	60	-
10	25	160	0.001203	10.7	175	110	100	60	-
11	20	80	0.003586	10.2	186	60	80	80	-
12	20	80	0.005513	9.9	230	40	80	80	[30 40] [55 65]
13	25	85	0.000371	13.1	225	30	80	80	-
14	15	55	0.001929	12.1	309	20	55	55	-
15	15	55	0.004447	12.4	323	20	55	55	-

B-co-efficients of 15-generating units

$$B_{tp,ij} = \begin{bmatrix}
0.0014 & 0.0012 & 0.007 & -0.0001 & -0.0003 & -0.0001 & -0.0001 & -0.0001 & -0.0003 & 0.0005 & -0.0003 & -0.0002 & -0.0004 & 0.0003 & -0.0001 \\
0.0012 & 0.0015 & 0.0013 & 0.0000 & -0.0005 & -0.0002 & 0.0000 & 0.0001 & -0.0002 & -0.0004 & -0.0004 & -0.0000 & 0.0004 & 0.0010 & -0.0002 \\
0.0007 & 0.0013 & 0.0076 & -0.0001 & -0.0013 & -0.0009 & -0.0001 & 0.0000 & -0.0008 & -0.0012 & -0.0017 & -0.0000 & -0.0026 & 0.0111 & -0.0028 \\
-0.0001 & 0.0000 & -0.0001 & 0.0034 & -0.0007 & -0.0004 & 0.0011 & 0.0050 & 0.0029 & 0.0032 & -0.0011 & -0.0000 & 0.0001 & 0.0001 & -0.0026 \\
-0.0003 & -0.0005 & -0.0013 & -0.0007 & 0.0090 & 0.0014 & -0.0003 & -0.0012 & -0.0010 & -0.0013 & 0.0007 & -0.0002 & -0.0002 & -0.0024 & -0.0003 \\
-0.0001 & -0.0002 & -0.0009 & -0.0004 & 0.0014 & 0.0016 & -0.0000 & -0.0006 & -0.0005 & -0.0008 & 0.0011 & -0.0001 & -0.0002 & -0.0017 & 0.0003 \\
-0.0001 & 0.0000 & -0.0001 & 0.0011 & -0.0003 & -0.0000 & 0.0015 & 0.0017 & 0.0015 & 0.0009 & -0.0005 & 0.0007 & -0.0000 & -0.0002 & -0.0008 \\
-0.0001 & 0.0001 & 0.0000 & 0.0050 & -0.0012 & -0.0006 & 0.0017 & 0.0168 & 0.0082 & 0.0079 & -0.0023 & -0.0036 & 0.0001 & 0.0005 & -0.0078 \\
-0.0003 & -0.0002 & -0.0008 & 0.0029 & -0.0010 & -0.0005 & 0.0015 & 0.0082 & 0.0129 & 0.0116 & -0.0021 & -0.0025 & 0.0007 & -0.0012 & -0.0072 \\
-0.0005 & -0.0004 & -0.0012 & 0.0032 & -0.0013 & -0.0008 & 0.0009 & 0.0079 & 0.0116 & 0.0200 & -0.0027 & -0.0034 & 0.0009 & -0.0011 & -0.0088 \\
-0.0003 & -0.0004 & -0.0017 & -0.0011 & 0.0007 & 0.0011 & -0.0005 & -0.0023 & -0.0021 & -0.0027 & 0.0140 & 0.0001 & 0.0004 & -0.0038 & 0.0168 \\
-0.0002 & -0.0000 & -0.0000 & -0.0000 & -0.0002 & -0.0001 & 0.0007 & -0.0036 & -0.0025 & -0.0034 & 0.0001 & 0.0054 & -0.0001 & -0.0004 & 0.0028 \\
0.0004 & 0.0004 & -0.0026 & 0.0001 & -0.0002 & -0.0002 & -0.0000 & 0.0001 & 0.0007 & 0.0009 & 0.0004 & -0.0001 & 0.0103 & -0.0101 & 0.0028 \\
0.0003 & 0.0010 & 0.0111 & 0.0001 & -0.0024 & -0.0017 & -0.0002 & 0.0005 & -0.0012 & -0.0011 & -0.0038 & -0.0004 & -0.0101 & 0.0578 & -0.0094 \\
0.0001 & -0.0002 & -0.0028 & -0.0026 & -0.0003 & 0.0003 & -0.0008 & -0.0078 & -0.0072 & -0.0088 & 0.0168 & 0.0028 & 0.0028 & -0.0094 & 0.1283
\end{bmatrix}$$

$$B_{p,io} = \begin{bmatrix} -0.0001 & -0.0002 & 0.0028 & -0.0001 & 0.0001 & -0.0003 & -0.0002 & -0.0002 & 0.0006 & 0.0039 & -0.0017 & -0.0000 & -0.0032 & 0.0067 & -0.0064 \end{bmatrix}$$

$$B_{tp,oo} = 0.0055$$

APPENDIX 5

TEST SYSTEM FOR CONVENTIONAL GENERATING UNITS (20-GENERATING UNITS WITHOUT VALVE POINT LOADING)

Table A5.1 Generating unit capacity and co-efficient

Units	$P_{tp,j}^{\min}$	$P_{tp,j}^{\max}$	$a_{tp,j}$	$b_{tp,j}$	$c_{tp,j}$
1	150	600	0.00068	18.19	1000
2	50	200	0.00071	19.26	970
3	50	200	0.00650	19.80	600
4	50	200	0.00500	19.10	700
5	50	200	0.00738	18.10	420
6	20	100	0.00612	19.26	360
7	25	125	0.00790	17.14	490
8	50	150	0.00813	18.92	660
9	50	200	0.00522	18.27	765
10	30	150	0.00572	18.92	770
11	100	300	0.00480	16.69	800
12	150	500	0.00310	16.76	970
13	40	160	0.00850	17.36	900
14	20	130	0.00511	18.70	700
15	25	185	0.00398	18.70	450
16	20	80	0.07120	14.26	370
17	30	85	0.00890	19.14	480
18	30	120	0.00713	18.92	680
19	40	120	0.00622	18.47	700
20	30	100	0.00773	19.79	850

B-co-efficients of 20-generating units

$$B_{\varphi,\vartheta} = 10^{-3}$$

```
18.70   0.43 − 4.61   0.36    0.32 − 0.66 0.96 −1.60 0.80 − 0.10 3.60 0.64 0.79         2.10    1.70 0.80 − 3.20 0.70            0.48    − 0.70  λ
0.43    8.30 − 0.97   0.22    0.75 − 0.28 5.04 1.70 0.54 7.20 − 0.28 0.98 − 0.46         1.30    0.80 − 0.20    0.52 − 1.70    0.80    0.20
− 4.61 − 0.97 9.00 − 2.00     0.63    3.00 1.70 − 4.30 3.10 − 2.00 0.70 − 0.77 0.93       4.60 − 0.30    4.20 0.38      0.70    − 2.00    3.60
0.36 0.22 − 2.00    5.30    0.47    2.62 −1.96 2.10 0.67 1.80 − 0.45 0.92 2.40           7.60 − 0.20    0.70 − 1.00    0.86    1.60    0.87
0.32    0.75    0.63    0.47    8.60 − 0.80    0.37    0.72 − 0.90    0.69    1.80 4.30 − 2.80 − 0.70    2.30    3.60    0.80    0.20 − 3.00    0.50
− 0.66 − 0.28    3.00 2.62 − 0.80    11.8 − 4.90 0.30 3.00 − 3.00 0.40 0.78 6.40 2.60 − 0.20 2.10 − 0.40           2.30    1.60    − 2.10
0.96 5.04    1.70 −1.96    0.37 − 4.90 8.24 − 0.90 5.90 − 0.60 8.50 − 0.83 7.20 4.80 − 0.90 − 0.10       1.30    0.76    1.90    1.30
−1.60    1.70 − 4.30    2.10 0.72    0.30 − 0.90 1.20 − 0.96 0.56 1.60       0.80 − 0.40 0.23 0.75 − 0.56       0.80 − 0.30    5.30    0.80
0.80    0.54 3.10    0.67 − 0.90    3.00 5.90 − 0.96 0.93 − 0.30 6.50       2.30 2.60 0.58 − 0.10       0.23 − 0.30 1.50    0.74    0.70
−0.10    7.20 − 2.00    1.80 0.69 − 3.00 − 0.60       0.56 − 0.30    0.99 − 6.60 3.90    2.30 − 0.30    2.80 − 0.80    0.38    1.90    0.47    − 0.26
3.60 − 0.28 0.70 − 0.45 1.80       0.40 8.50       1.60 6.50 − 6.60       10.7 5.30 − 0.60 0.70    1.90 − 2.60    0.93 − 0.60    3.80    −1.50
0.64    0.98 − 0.77    0.92 4.30    0.78 − 0.83 0.80 2.30 3.90       5.30 8.00    0.90 2.10 − 0.70 5.70    5.40 1.50    0.70    0.10
0.79 −0.46 0.93    2.40 − 2.80    6.40 7.20 −0.40 2.60 2.30 −0.60 0.90       11.0 0.87 −1.00 3.60    0.06 − 0.90 0.60    1.50
2.10    1.30 4.60    7.60 − 0.70    2.60 4.80 0.23 0.58 −0.30 0.70 2.10       0.87 3.80 0.50 −0.70    1.90 2.30 − 0.97    0.90
1.70    0.80 − 0.30 −0.20 2.30 −0.20 −0.90 0.75 −0.10 2.80 1.90 − 0.70 −1.00 0.50    11.0    1.90 − 0.80 2.60    2.30    −0.10
0.80 −0.20 4.20 0.70 3.60 2.10 −0.10 − 0.56 0.23 − 0.80 − 2.60 5.70       3.60 − 0.70    1.90    10.8 2.50 −1.80    0.90    − 2.60
−3.20 0.52    0.38 −1.00 0.80 −0.40    1.30    0.80 −0.30    0.38    0.93 5.40    0.46    1.90 − 0.80 2.50    8.70    4.20 − 0.30    0.68
0.70 −1.70 0.70    0.86 0.20    2.30    0.76 −0.30 1.50    1.90 − 0.60 1.50 −0.90 2.30    2.60 −1.80 4.20    2.20    0.16    −0.30
0.48    0.80 − 2.00    1.60 −3.00    1.60    1.90 5.30 0.74    0.47 3.80 0.70 0.60 −0.97    2.30 0.90 −0.30    0.16    7.60    0.69
λ −0.70    0.20    3.60    0.87 0.50 − 2.10    1.30    0.80 0.70 −0.26 −1.50 0.10    1.50 0.90 −0.10 −2.60    0.68 −0.30    0.69    7.00  ϑ
```

APPENDIX 6

TEST SYSTEM OF WIND FARM AND SINGLE WIND TURBINE

Table A6.1 wind farm and wind turbine data

Rated power	2MW
Scale factor	6 meter/second
Shape factor	2 meter/second
Cut-in speed	2.5 meter/second
Cut-off speed	15 meter/second
Radius of the turbine	30 meter
Multiplication factor of gearbox	60
Air density	1.225 kgm-3
Number of poles	4
Nominal voltage	960V
HVDC voltage	150kv
Rating of the transformer	2.5MVA
Wind farm transformer rating	100MVA

REFERENCES

1. Andrew, K & Zhe, S 2010, 'Design of wind farm layout for maximum wind energy capture', Renewable energy, vol. 35, pp. 685-694.

2. Aniruddha, B & Chattopadhyay, P 2010, 'Biogeoraphy-based optimization for different economic load dispatch problems', IEEE Transaction on power systems, vol. 25, no. 2, pp. 1064-1077.

3. Aruldoss Albert Victoire, T & Ebenezer Jeyakumar 2004, 'Hybrid PSO–SQP for economic dispatch with valve-point effect', Electric Power Systems Research , vol. 71, pp. 51-59.

4. Bansal, R 2005, 'Optimization methods of electric power systems: An overview', International Journal of Emerging Electric Power System, vol. 2, no. 1, pp. 1-9.

5. Basu, M 2015, 'Modified Particle Swarm Optimization for non-convex economic dispatch problems', Electrical power and energy systems, vol. 69, pp. 304-312.

6. Basu, M & Chowdhury, A 2013, 'Cuckoo Search algorithm for economic dispatch', Energy, pp. 1-10.

7. Benhamida, F, Salhi, Y, Souag, S, Graa, A , Ramdani, Y & Bendaoud, A 2013, 'A PSO algorithm for economic scheduling of power system incorporating wind based generation', 5th International Conference on Modeling, Simulation and Applied Optimization (ICMSAO) , pp. 1-6.

8. Bergh, VF, Andries, P & Engelbrecht 2004, 'A Cooperative Approach to Particle Swarm Optimization', IEEE Transaction on Evolutionary computation, vol. 8, no. 3, pp. 225-240.

9. Beyer, HG & Deb, K 2001, 'On self-adaptive feature in real parameter evolution algorithm', IEEE Transaction on evolutionary computation, vol. 5, no. 3, pp. 250-568.

10. Binetti, G, Ali, Frank, L, Lewis, David & Biagio 2014, 'Distributed Consensus-based economic dispatch with transmission losses', IEEE Transactions on Power systems, vol. 29, no. 4, pp. 1711-1720.

11. Cheng-Chien Kuo 2008, 'A noval coding scheme for practical economic dispatch by modified particle swarm approach', IEEE Transaction on Power Systems, vol. 23, no. 4, pp. 1825-1835.

12. Chiang, C L 2005, 'Improved Genetic algorithm for power economic dispatch of units with valve point effects and multiple fuels', IEEE Transacation on power systems, vol. 20, no. 4, pp. 1690-1699.

13. Deb, K & Agarwal, RB 1995, 'Simulated binary cross over for continous search space', Complex system, vol. 9, pp. 115-148.

14. Deb, K & Goyal, M 1996, 'A combained genetic adaptive search (GeneAS) engineering design', Comuter science congress, vol. 26, no. 4, pp. 30-45.

15. Elfyn Hughes, R 1958, 'Sheep population and environment in snowdonia', Journal of ecology, vol. 46, no. 1, pp. 169-189.

16. EL-Khazendar, MK & Ahmed, MM 1994, 'A generating wind scheme for maximum energy capture and minimum cost', Renewable Energy, vol. 5, no. 1, pp. 709-711.

17. Fan, HY & Lampinen, J 2003, 'A Trigonometric Mutation Operation to Differential Evolution', Journal of Global Optimization, vol. 27, no. 1, pp. 105-129.

18. Fan, JY & McDonald, JD 1994, 'A practical approach to real time economic dispatch considering unit's prohibited operating zones', IEEE Transaction on Power Systems, vol. 9, no. 4, pp. 1737-1743.

19. Farhat, IA & El-Hawary, ME 2010, 'Dynamic adaptive bacterial forging algorithm for optimum economic dispatch with vale-point effects and wind power', IET Generation, Transmission & Distribution, vol. 4, no. 9, pp. 989-999.

20. Gaing, ZL 2003, 'Particle Swarm Optimization to Solving the Economic Dispatch Considering the Generator Constraints', IEEE transactions on power systems, vol. 18, no. 3, pp. 1187-1195.

21. Gebraad, P & Wingerden, JWV 2014, 'Maximum power-point tracking control for wind farms', Wind energy, vol. 18, no. 3, pp. 429-447.

22. Gillella, S & Geetha, V 2008, 'A modified particle swarm optimizaiton to solve the economic dispatch problem of thermal generators of power system', Journal of electrical engineering, pp. 74-84.

23. Giulio Binetti, Ali, D, Frank, L, Lewis, L, David Naso & Biagio, T 2014, 'Distributed Consensus-Based Economic Dispatch With Transmission Losses', IEEE Transaction on power systems, vol. 29, no. 4, pp. 1711-1720.

24. Glare, PGW 1990, The Oxford Latin Dictionary.

25. Goldberg, DE 1989, Genetic Algorithms in search, optimization & machine learning, Addison--Wesley, MA.

26. Gong, Y, Hodgson, J, Lambert, MG & Gordon, IL 1996, 'Short term ingestive behaviour of sheep and goats grazing grasses and legumes', Newzealand Journal of agricultural research, vol. 39, pp. 63-73.

27. Helen Armstrong 1993, The grazing behaviour of large herbivores in the uplands.

28. Hemamalini, S, Sishaj, P & Simon 2010, 'Artificial bee colony algorithm for economic load dispatch problem with non-smooth cost function', Electric power components and systems, vol. 38, pp. 786-803.

29. Hetzer, JD & Bhattarai, K 2008, 'An economic dispatch model incorporating wind power', IEEE transaction on energy conversion, vol. 23, no. 2, pp. 603-611.

30. Hui, Li, McLaren, P & Shi, KL 2005, Neural network based sensorless maximum wind energy capture with compensated power coefficient : Industry Applications Conference , pp. 1548- 1556.

31. Immanuel S & Thanushkodi K 2007, 'A New Particle Swarm Optimization Solution to Nonconvex Economic Dispatch Problems', IEEE transactions on power systems, vol. 22, no. 1, pp. 42-51.

32. Immanuel Selvakumar, A & Thanushkodi, K 2008, 'A new particle swarm optimiztion solution to non-convex economic dispatch problems', IEEE Transaction on Power Systems, vol. 78, pp. 2-10.

33. Immanuel Selvakumar, A & Thanushkodi, K 2009, 'Optimization using civilized swarm:solution to economic dispatch mutiple minima', Electrical Power Systesms Research, vol. 79, no. 1, pp. 8-16.

34. James, JQ, Yu & Victor, OKL 2016, 'A social spider algorithm for solving the non-convex economic load dispatch problem', Neurocomputing, vol. 171, pp. 955-965.

35. Jiang, W, Yan, Z & Hu, Z 2011, 'A Novel Improved Particle Swarm Optimization Approach for Dynamic Economic Dispatch Incorporating Wind Powe', Electric Power Components and Systems, vol. 39, no. 5, pp. 461-477.

36. Kennedy, J & Eberhart, R 1995, Particle swarm optimization : Neural Networks , pp. 1942-1948.

37. Kmath. 2014. The Mystry of the grazing goat. Available from : <www.mathpages.com/home/kmath074/kmath074.htm>

38. Kothari, D & Dhilon, JS 2012, Power System Optimization, PHI learing Pvt Ltd.

39. Leandro dos & Lee, C 2008, 'Solving economic load dispatch problems in power systems using chaotic and gaussian particle swarm optimization', Electrical power and energy systems, vol. 30, pp. 297-307.

40. Lee, FN & Breipohl, AM 1993, 'Reserve contrained economic dispathc prohibited operating zones', IEEE Transaction on power systems, vol. 8, no. 1, pp. 246-254.

41. Lee, JC, Lin, WM, Liao, GC & Tsao, P 2011, 'Quantum genetic algorithm for dynamic economic dispatch with valve-point effects and including wind power system', Electrical power and energy systems, vol. 33, no. 2, pp. 189-197.

42. Ling Wang & Ling-po Li 2013, 'An effective differential harmony search algorithm for the solving non-convex economic load dispatch problems', Electrical Power and Energy Systems, vol. 44, pp. 832-843.

43. Lin, CE & Viviani, GL 1984, 'Hierarchical economic dispatch for piecewise quadratic cost functions', IEEE Transactions on Power Apparatus and Systems, vol. 103, no. 6, pp. 1170-1175.

44. Mahshid, HM & Mehdi, GD 2013, 'Economic Dispatch Incorporating Wind Power Plant Using Modified Particle Swarm Optimization', Journal of Recent Sciences, vol. 2, no. 6, pp. 108-112.

45. Mallipeddi, R, Suganthan, PN, Pan, Q & Tasgetiren, MF 2011, 'Differential evolutionalgorithm with ensemble of parameters and mutation trategies', Applied Soft Computing , vol. 11, no. 2, pp. 1679-1696.

46. Manwell, JF, McGowan, JC & Rogers, AL 2009, 'Wind energy explained theory design and application', IEEE Transaction on Smart Grid, vol. 1, no. 3, pp. 347-355.

47. Miranda, V & Hang, PS 2005, 'Economic Dispatch Model With Fuzzy Wind Constraints and Attitudes of Dispatchers', IEEE Transaction on power systems, vol. 20, no. 4, pp. 2143-2145.

48. Mirhamed Mola & Faridoon Shabaninia 2013, 'Design and simulation of type II fuzzy logic controller for capturing maximum wind energy', International Journal of Ambient Energy, vol. 35, no. 3, pp. 110-117.

49. MNRE 2017, All India Installed Capacity (in mw) of Power Stations.

50. Mukund R Patel 2006, Wind and Solar power systems Design, Analysis and operation, Taylor & Fransi.

51. Muthu, VPS & Thanushkodi, K 2012, 'Considering transmission loss for an economic dispatch problem without valve-point loading using an EP-EPSO algorithm', Turkish Journal of Electrical Electrical Engineering and computer science, vol. 20, no. 2, pp. 1259-1267.

52. Naresh, R & Dubey, J 2004, 'Two-phase neural network based modelling framework constrained economic load dispatch', Generation transmission Distriution, vol. 151, no. 3, pp. 373-378.

53. Nasimul, N & Hitoshi Iba 2008, 'Differential evolution for economic load dispatch problems', Electric power system research, vol. 78, pp. 1322-1331.

54. Nicholson, IA 1970, 'Some effects of animal grazing and browsing on vegetation', botanical journal of scotland, vol. 58, no. 2, pp. 211-219.

55. Nidul Sinha, Chakrabarti, R & Chattopadhyay, P 2003, 'Evolutionary Programming techniques for economic load dispatch', IEEE Transactions on Evolutionary computation, vol. 7, no. 1, pp. 83-94.

56. Nima, A & Hossein, S 2010, 'Soltuon of non-convex economic dispatch problem considering valve point effect by a new modified differential evolution algorithm', Electrical Power and Energy Systems, vol. 32, pp. 893-903.

57. Oriol Gomis-Bellmunt, Adria Junyen-Ferre, Andreas Sumpe & Samuel Galceran-Arellano 2010, 'Maximum generation power evaluation of variable frequency off shore wind farms when connected to a single powerconverter', Applied energy, vol. 87, pp. 3103-3109.

58. Park, JH, Kim, YS, Eom, IK & Lee, KY 1993, 'Economic load dispatch for piecewise quadratic cost function using hopfield neural network', IEEE Transaction on power systems, vol.8, no. 3, pp. 1030-1038.

59. Pedersen, M 2010, Good Parameters for Differential Evolution.

60. Pereira-Neto, A, Unsihuay C & Saavedra, O 2005, 'Efficieint evolutionary strategy optimization procedure to solve the nonconvex economic dispatch problem with generator constraints', IEEE Proceding generaton transmission and distribution, ed. IEEE, pp. 653-660.

61. Po-Hung, Hong-Chan & Chan, G 1995, 'Large-Scale Economic Dispatch by Genetic Algorithm', IEEE Transactions o n Power Systems, vol. 10, no. 4, pp. 1919-1927.

62. Rabih, AJ, Alun, HC & Brian, JC 2000, 'A Study of the Homogeneous Algorithm for Dynamic Economic Dispatch with Network Constraints and Transmission Losses', IEEE Transaction on Power Systems, vol. 15, no. 2, pp. 605-611.

63. Ratnaweera, A, Halgamuge, SK & Watson, HC 2004, 'Self orgaanizing hierarchical particle swarm optimizer with time varying acceleration coefficients', IEEE Transaction on Evolutionary computation, vol. 8, no. 3, pp. 240-255.

64. Roy, P, Ghoshal, S & Thakur, S 2014, 'Biogeogaphy-based optimization for economic load dispatch', ELectric power components and systems, vol. 38, no. 2, pp. 166-181.

65. Rozenn, Michael, C, Torben, L & Uwe, P 2010, Simulation of shear and turbulance impact on wind turbine performance. Available from : <http://130.226.56.153/ripubl/reports/ris-r-1722.pdf>

66. Sanjay Roy 2012, 'Inclusion of short duration wind variations in economic load dispatch', IEEE transaction on sustainable energy, vol. 3, no. 2, pp. 265-273.

67. Sayyid, MS, Ahmad, SY & Farzad, F 2012, 'A new gumption approach for economic dispatch problems with losses effect based on valve point active power', Electrical Power System Research, vol. 92, pp. 81-86.

68. Shahryar, R, Hamid, RT & Salama, MMA 2008, 'Opposition-Based Differential Evolution', IEEE Transaction on evolutionary computation, vol. 12, no. 1, pp. 64-82.

69. Shi, Y & Eberhart, R 1998, A modified particle swarm optimizer : International conference on evolutionary computaion , pp. 69-73.

70. Shuhui Li , Timothy, A, Haskew, A & Yang-Ki H 2011, 'Investigation of maximum wind power extraction using adaptive virtual lookup-table approach ', Energy research, vol. 35, pp. 964-978.

71. Souma, C, Jie, Z, Achille, M & Luciano, C 2012, 'Unrestricted wind farm layout optimization (UWFLO):Investigating key factors influencing the maximum power generation', Renewable energy, vol. 38, pp. 16-30.

72. Srinivasa Reddy, A & Vaisakh, K 2013, 'Shuffled differential evolution for large scale economic dispatch', Electrical Power Systems Research, vol. 96, pp. 237-245.

73. Storn, R & Price, K 1997, 'Differential evolution-A Simple and Effcient Heuristic for Gloabal Optimization over Contiinous Space', Journal of Global Optimization, vol. 11, pp. 341-359.

74. Subbaraj, P & Rajnarayanan, PN 2009, 'Optimal reactive power dispatch using self-adaptive real coded genetic algorithm', Electrical Power Systems Research, vol. 79, pp. 374-381.

75. Subbaraj, P, Rengaraj, R & Salivahanan, S 2009, Real-Coded genetic algorithm enhanced with self adaptation for solving economic dispatch problem with prohibited zone : International conference on control, automiation, and enenergy conservation , pp. 1-6.

76. Subbaraj, P, Rengaraj, R & Salivahanan, S 2011, 'Enchancement of self-adaptive real-coded genetic algorithm using Taguchi method for economic dispatch problem', Applied Soft Computing, vol. 11, pp. 83-92.

77. Subbaraj, P, Rengaraj, R, Salivahanan, S & Senthilkumar, T 2010, 'Parallel particle swarm optimization with modified stochastic acceleration factors for solving large scale economic dispatch problem', Electrical Power and Energy Systems, vol. 32, pp. 1014-1023.

78. Su, CT & Lin, CT 2000, 'New approach with hopefield modelling framework to economic disapatch problem', IEEE Transaction on Power System , vol. 15, no. 2, pp. 541-545.

79. Thang, T & Dieu, N 2015, 'The application of one rank cuckoo search algorithm for solving economic load dispatch problems', Applied Soft Computing, vol. 37, pp. 763-773.

80. Walters, DC & Sheble, GB 1993, 'Genetic algorithm solution of economic dispatch with valve point loading', IEEE Transaction on power system, vol. 8, no. 3, pp. 1325-1331.

81. Wong, KP, Lau, BS & Fry, A 1993, Modelling generator input output characteristics with valve point loading using neural networks: International conference on advanced power system control operation and management , pp. 843-848.

82. Wood, AJ & Wollenberg, B 1996, Power Generation, Operation, and control, Wiley, New York.

83. Wood, AJ, Wollenberg, BF & Sheble, GB 1996, Power Generation, Operation, and control, Wiley, New York.

84. Yalcionoz, T, Altum, H & Uzam, M 2001, Economic dispatch solution using a genetic algorithm based arithmetic cross over : International conference on Proto Power Tech.

85. Yang, HT, Yang, PC & Huang, CL 1996, 'Evolutionary programming based economic dispatch for units with non-smooth fuel cost function', IEEE Transaction on Power Systems, vol. 11, no. 1, pp. 112-118.

86. Yingming Liu, Enyong, Y, Xiaodong, W & Hongfang, X 2013, 'Research on capture the maximum Wind Energy Base on fuzzy PID controller', Applied Mechanics and Materials, vol. 268, pp. 1422-1428.

87. Zhang, J, Cheng, M, Chen, Z & Fu, X 2008, Pitch Angle Control for Variable Speed Wind Turbines : DRPT, pp. 2691-2696.

88. Zhao, B, Guo, CX & Cao, YJ 2005, 'A multiagent-based particle swarm optimization approach for optimal reactive power dispatch', IEEE Transaction on Power Systems, vol. 20, no. 2, pp. 1070 - 1078.

89. Zivkovic, M, Vlastimir, D, Gradimir, S & Ivan, T 2012, 'Hybrid soft computing control strategies for improving the energy capture of a wind farm', Thermal science, vol. 16, no. 2, pp. 483-491.

www.ingramcontent.com/pod-product-compliance
Lightning Source LLC
Chambersburg PA
CBHW081147130726
47996CB00009B/3028